NAPOLEONIC WARFARE

NAPOLEONIC WARFARE

The Operational Art of the Great Campaigns

JOHN T. KUEHN

PRAEGER™

An Imprint of ABC-CLIO, LLC
Santa Barbara, California • Denver, Colorado

Library of Congress Cataloging-in-Publication Data

Kuehn, John T.
Napoleonic warfare : the operational art of the great campaigns / John T. Kuehn.
pages cm
Includes bibliographical references and index.
ISBN 978-1-4408-3307-6 (alk. paper) — ISBN 978-1-4408-3308-3 (EISBN)
1. Napoleonic Wars, 1800–1815—Campaigns. 2. Operational art (Military science)—History—19th century. 3. First Coalition, War of the, 1792–1797—Campaigns. 4. Second Coalition, War of the, 1798–1801—Campaigns. I. Title.
DC226.4.K84 2015
940.2'74—dc23 2015005479

ISBN: 978-1-4408-3307-6
EISBN: 978-1-4408-3308-3

19 18 17 16 15 1 2 3 4 5

This book is also available on the World Wide Web as an eBook.
Visit www.abc-clio.com for details.

Praeger
An Imprint of ABC-CLIO, LLC

ABC-CLIO, LLC
130 Cremona Drive, P.O. Box 1911
Santa Barbara, California 93116-1911

This book is printed on acid-free paper ∞

Manufactured in the United States of America

CONTENTS

MAPS AND FIGURES

MAPS

FIGURES

PREFACE

WAR IS THREE DIMENSIONAL

This book has a very simple premise: European warfare from 1793 to 1815 lends itself particularly well to an operational-level analysis. Modern war is three-dimensional, divided by modern doctrine into three "levels of war."[1] At the lowest end of the spectrum is the tactical level, the level of the engagement or the battle. Combat. At the high end are strategy and grand strategy, how the military forces are used with other forms of national power to achieve political ends. In between these two is the operational level, the level of the campaign, or a series of engagements, fought to achieve a particular military purpose, but which might or might not lead to a situation resulting in final military victory or a suspension of hostilities.

It has only been in modern times that the term *operational level of war* has come to encompass its own distinct niche in discussions of military art and science.[2] However, giving a phenomenon or concept a name does not presuppose the nonexistence of that phenomenon prior to introduction of its particular usage. Such is the case with the operational level of war. This term is of recent vintage and identifying when it came into common use depends on which professional military and its doctrine one is examining. I learned of it in my teenage years, until then being unaware of anything beyond battles and tactics at one end and war and strategy at the other.[3] Most military historians treated the topic similarly (and some still do). War, it seemed, was naturally two-dimensional—tactics and strategy. War is certainly not two-dimensional, nor is it linear. Of the three levels of war—strategic, operational, and tactical—the one that is least understood and written about resides in that always uncomfortable middle ground, the operational level.

The size, scope, and complexity of operations during the Napoleonic era increased exponentially along with the profound political events shaking and shaping the societies of Europe. Huge armies, then army groups, and finally entire armed societies "went at it" in a series of complex and interlinking campaigns on a scale not seen in Europe since Roman times. Carl von Clausewitz, arguably among the first of the modern operational analysts, famously wrote: "but in 1793, a force appeared that beggared all imagination. Suddenly war again became the business of the people—a people of thirty millions, all of whom considered themselves to be citizens."[4] He wrote about France and the force of armed nationalism that the French Revolution unleashed upon Europe. Armed nationalism is not a thing of the past and neither is revolution. In the past we find the stuff of the present, patterns that, if examined, can help us navigate the troubled waters of our own day.

Clausewitz's judgment captures one of more profound truths about the military revolution that occurred during the wars that took place after the French Revolution began.[5] This force that "beggared all imagination" was also a force of exceptional complexity—as complex as the nation itself. And we must remember that France was at the forefront of the other nations of Europe in terms of social complexity. New systems, methods, tactics, doctrines, and organizations were needed to enable this juggernaut to operate. There was a revolution in the complexity of war. New leadership was needed, and it was in this environment that "new men" from the minor nobility and middle classes like Napoleon Bonaparte and Gerhard von Scharnhorst prospered. Additionally, enlightened leaders from the status quo regimes such as the Austrian archduke Charles, Sir Arthur Wellesley (later the Duke of Wellington), and Marshal A. A. Suvorov emerged as innovators from the more traditional militaries of the monarchical enemies of France. They came to grips with the complexity they faced and began to wage war in a new way that matched the complexity of operations demanded by the challenge from France and its armed nationalism empowered by the ideology of the "Rights of Man."[6]

Given the complexity of warfare during the period 1792–1815, it lends itself naturally to an operational-level investigation. Accordingly, the character of this book will reflect two operational frameworks of analysis. The first is that developed by the first generation of Soviet military thinkers between 1918 and 1941. The Germans had pioneered the usage and vocabulary for an operational level of war. However, as with most new concepts, it was not the discoverers who established the mature framework for a coherent approach to an operational approach to practicing war at this level. The Soviets reaped where the Germans had sown.[7] The second approach is that of James Schneider, formerly of the School of Advanced Military Studies (SAMS) of the U.S. Army Command and General Staff College. Schneider uses the language of fluid dynamics and physics to describe the components of operational art.[8]

Most of the sources used for this book are secondary and may appear dated with respect to their interpretations. They are used to get the basic facts, which at the operational level of war are largely established—for example, it is clear that the Napoleon was victorious in the Austerlitz campaign while he was not in the 1812 Russian campaign. Some of the conclusions presented here are not startlingly new as far as the facts are concerned; research using new or little-used primary resources is necessary for a scholar to make such a claim. This book instead proposes a new line of analysis for the Napoleonic Wars along principally operational lines. I also realize that using 20th-century military theory to explain 18th and 19th century events has some inherent analytical flaws, superimposing concepts that were not part of the framework that Napoleonic leaders and generals worked from in making military decisions. However, it can be argued (and will be later in this book) that at the very least Napoleon and his talented subordinate Baron Antoine de Jomini understood the basic principles of operational war and, at least in Jomini's case, had codified the first cut. The Prusso-Germans, too, seemed to understand the profound implications of the emergence of a new level of war, and Clausewitz even referred to them as "operative elements" in *On War* in referencing his book that deals with military "strategy" at the campaign level.[9]

The book will focus primarily at the campaign level and Napoleonic tactics will not be explored in detail and only come up in discussion as they directly relate to the operational level—such as the use of cannon by brigades and regiments of the period to give them more firepower and thus operational durability. For the reader who wants more on tactical developments of the period, David Chandler's fine discussion entitled "Napoleon's Art of War" in *The Campaigns of Napoleon* has some excellent discussions of tactical evolutions during the period, especially on the issues of skirmishing and various tactical formations such as the column, line, and mixed order. For even more detail, Brent Nosworthy's *With Musket, Cannon and Sword* synthesizes all the elements of the battlefield to give one an accurate view inside the "black box" of the tactics as they were actually executed on the battlefields of the period.[10]

In summary, this book aims for something of an impressionistic result that suggests the operational-level approach adopted here illustrates effectively the more esoteric concept of operational art—how military genius, as best defined by Clausewitz, operated in space and time in the uncertain environment at the operational level during the era of a veritable "God of War."[11]

ACKNOWLEDGMENTS

I owe a debt of gratitude to David Chandler for his wonderful *Campaigns of Napoleon* that first inspired my interest in the Napoleonic era. Professor Jonathan M. House, my longtime friend and mentor, spent tireless hours reviewing the manuscript and doing his usual great job of editing my turgid prose. Jon is national treasure and should be recognized as such. I also want to thank both Tom Huber and Bob Epstein, who mentored me when I first wrote on these topics back in the 1990s while assigned to Fort Leavenworth. Steve Catalano of Praeger was also instrumental in his support and encouragement for my proposal. Finally, I am most grateful to my partner, friend, and wife Kimberlee for her untiring support throughout the entire project. It goes without saying that all errors are mine and mine alone.

ONE

OPERATIONAL ART

All that pertains to operations as such takes place, on the whole, independently of actual combat, whereas in the term "strategisch" (strategical) things become easily confused, as has been proved by the example of our enemies who are wont to speak of strategical conditions when it is merely a question of purely local matters. At any rate, the term "strategy" ought to be confined to the most important measures of high command.

General Hugo von Freytag-Loringhoven[1]

What is operational art and where did it come from? Most simply, operational art is the design and conduct of campaigns just as tactics are the design and conduct of battles and engagements. Operational art is normally employed by senior military leaders using what we will refer to as *operational-level units*—these types of units can act independently or in concert with each other over a sustained period of time. On land usually an army corps or higher can be considered an operational unit, although sometimes divisions can be characterized as such, but an operational unit usually does not encompass the totality of a nation's military land power. At sea it is a bit different, but the fundamental rule gives us a grouping of warships referred to variously as an independent squadron or fleet, but normally not the totality of the nation's, or coalition's, naval forces.

As to the origins of operational art, the first serious writing on it and on the operational level of war can be identified in the 20th century by Imperial Russian and later Soviet military officers. These officers built on ideas that Baron Antoine Jomini, Carl von Clausewitz and, later, the military historian Hans Delbrück discussed in their various writings.[2] It might seem strange to start out a book on the Napoleonic Wars with a discussion that includes Soviet military theory, but it makes sense when one remembers that often we humans make sense of things after they have occurred and not while they are occurring. This was the case for the political-strategic-military revolution wrought by the French Revolution.[3]

THE "DISCOVERY" OF THE OPERATIONAL LEVEL OF WAR

Napoleon's two contemporary interpreters, Jomini and Clausewitz, are the intellectual discoverers of what we today understand as the operational level of war. Antoine Jomini was first. The son of a Swiss banker, from the middle class of that day, the developments of the French Revolution fascinated him. He also had a thirst for military history. He conducted what might be called an operational study of the campaigns of Frederick the Great. From that intellectual endeavor he came up with his famous "principles of war," embodied in the concepts of *maneuver*, *mass*, *offensive*, and *objective*. It was only after this initial study that Jomini came to Napoleon Bonaparte's attention, as well as coming to see the reflection of his principles in the style of war that Napoleon practiced.[4]

As Professor Tom Huber has said, "Jomini created a theory of deployments," or for our purposes a theory of how to integrate logistics and maneuver into a theory of *operational* principles.[5] The durability of these principles can be confirmed easily by looking at the recent doctrine of the U.S. Army in Field Manual (FM) 3–0 *Operations*. More recent U.S. doctrines such as Army Doctrinal Publication 3–0 and Army Doctrinal Reference Publication 3–0 have deleted a listing of these principles, but the careful reader can still find them in the sections on "operational art" and "the tenets of unified land operations."[6]

At the same time, the Prussian officer Carl von Clausewitz was also reflecting on the nature of war, although publication of his work would not occur until after his death in 1831. Clausewitz also gained his inspiration from the French and every page of his masterpiece *On War* breathes that inspiration. He spread his ideas about operational art throughout his work, but his most formal notions and writing about operational art can be found principally in three "books" from *On War*: Book Two on theory, Book Three on strategy, and Book Six, his lengthiest, on defense. In Book Two, he discusses war as both a science and an art, and clearly makes room for what he calls "creative ability."[7] Clausewitz devotes the entirety of Book Three to a discussion of strategy and first defines it as:

> Strategy is the use of the engagement for the purpose of the war. The strategist must therefore define an aim for the *entire operational side of the war* that will be in accordance with its purpose. . . . he will draft the plan of the war, and the aim of the war will determine the series of actions intended to achieve it: he will, in fact, shape the individual campaigns and, within these, decide on the individual engagements.[8] [emphasis mine]

Compare this with the Army definition found in Field Manual 3–0:

> The operational level links employing tactical forces to achieving the strategic end state. At the operational level, commanders conduct *campaigns* and major operations to establish conditions that define that end state.[9] [emphasis mine]

Clearly Clausewitz's definition of strategy describes the operational level of war. The context of Clausewitz's times, the Napoleonic Wars, is betrayed by where he locates the strategist in the activity of war—not back in the capital but, "it follows that the strategist must go on campaign himself."[10] The strategist is a campaigner in the early 19th century, but not so in the 20th and 21st. As one proceeds through Book Three's discussion, one finds just how congruent Clausewitz's view of "strategy" in action was with that of his contemporary Baron Antoine Jomini. The chapters in Book Three sound like a Jominian recitation—"Surprise, Cunning, Economy of Force, The Geometrical Factor."[11] Clausewitz's book on strategy is clearly about the operational level of war. He does cloud the issue slightly in linking war aims to the campaign (which really is a function of strategy), but effective operational art does not operate in a vacuum but serves the ends of strategy.

THE SOVIET OPERATIONAL ART MODEL

Clausewitz and Jomini's approach to operations and operational art has sometimes been called linear. Jomini, in particular, talked about lines of operations, but Clausewitz did as well, although both emphasized that a single line of operation no longer needed to exist to approach an objective in war, but that these did tend to conform to geography and established lines of communications such as roads and rivers. However, it was the Russians who developed some of Clausewitz's ideas a bit further, once removed by way of another German, the historian Hans Delbrück.

Hans Delbrück's great contribution to the theory of war was to break strategy into two types, an annihilative strategy (*Neiderwehrfungsstrategie*) that tended toward quick, rapier-like offensives and an attrition strategy of exhaustion (*Ermattungsstrategie*) that emphasized, often, the defense, and wearing one's enemy down until the final offensive coup de grâce. Delbrück used the same database as Jomini, the campaigns of Frederick the Great, and found that Frederick triumphed in his war more through the use of exhaustive strategies and maneuver than through the annihilation of his enemy in short decisive campaigns like Napoleon conducted for most of his career.[12]

These findings did not endear Delbrück to the German officer corps of his day. Among the reasons the Germans discovered but did not develop further the operational approach (at least in their officer corps) had as much to do with the successful practice of a sort of operational art 1.0 (using software nomenclature) by the elder Helmut von Moltke and his contemporaries. Success in two relatively short wars—the Austro-Prussian War (1866) and the Franco-Prussian War (1870–1871) seemed to indicate no real pressing problems that needed to be solved with the existing German "way of war." Alfred von Schlieffen's two-front nightmare of fighting the French and Russians was to be solved with Jominian operational approaches with a dash of Clausewitz, not with protracted campaigns that might hinge on the defense.[13]

For the leaders of the Imperial Russian Army, the German intellectual framework formed the starting point for their thinking about how to organize, mobilize, and (most of all) think about war in the brave new industrial world. Their defeat in the Crimean War against the Turkish and Anglo-French forces convinced them of the need for reform and they turned deliberately to a Prussian model that included freeing their serfs (as the Prussians had done after their defeat in 1807). Freed serfs provided the Russians a huge manpower resource for conscription on the Prussian model, not just a reflection of Western liberal principles and the "rights of man." These reforms also resulted in a German system for operational staffs and military education that would prove breeding grounds for the nascent operational artists and theorists of both the late Tsarist and early Red armies.[14] Nevertheless, the Russians made substantial intellectual progress of their own, but initially within the confines of a German operational design that depended on the short decisive war of maneuver without fully understanding how Moltke's approach differed from the Jominian-Napoleonic models.[15] However, their experience in the Russo-Japanese War—where they had exhausted the Japanese Army despite numerous errors—and preparation for a war with both the German and the Austrian-Hungarian Empires led to certain Russian staff officers to think about industrial war at the operational level under protracted circumstances. "In a word, Jomini was out, and Clausewitz and the elder Moltke (as modified by experience and observation) were in." Russian officers such as A.A. Nezmanov and especially N.P. Mikhnevich examined a new level of war between tactics and strategy at the General Staff Academy. Their views, in particular, emphasized the importance of operational planning, either for short war or for protracted war.[16]

However, it was the experience of World War I that finally pushed the Russian and then Soviet officer corps to begin to put together a framework for what became mature operational art in the 20th century. A.A. Svechin, writing about strategy after the war, essentially updated the framework proposed by Delbrück for the unique challenges (or one might say perceived threats) and geography of the new Soviet

State.[17] War must be a tapestry of campaigns, depending on circumstances, that would either be offensive or defensive in character. Given the exhaustion of the Soviet Union after constant war from 1914 to 1922, and the need to rebuild an industrial economy, Svechin naturally favored a strategic defense and exhaustion as best suiting short-term Soviet needs, although he also believed in the offensive under the right circumstances based on his study of the famous Brusilov Offensive in 1916. In his 1926 book, Svechin for the first time defined operational art: "Then, battle is the means of the operation. Tactics are the material of operational art. The operation is the means of strategy, and operational art is the material of strategy. This is the essence of the three-part formula given above."[18] Svechin's ideas irritated the inclinations of Marshal Mikhail Tukachevsky, who believed that the Soviet Union, if properly mobilized and correctly fighting in depth on the battlefield, could win with a series of operational blows on the offensive "as the means to conduct decisive operations in a total war."[19]

It was the writing and thinking of those considering how to construct and synchronize successive engagements and campaigns that the operational art "Soviet-style" came into being. Studying World War I, the Soviets, and especially Svechin, identified in the Brusilov Offensive during the summer of 1916 the components for offensive success, despite the ultimate defeat of that operation later that year. Brusilov had solved the problem of the static front with intelligent use of artillery and infiltration tactics, but his offensive had ultimately failed, at least as analyzed after the war by Soviet officers, because it had advanced on too many fronts without setting aside any reserves to exploit the breakthroughs in the front.[20] With the problem of how to restore maneuver to the battlefield solved, the Soviets turned their attention to how to sustain such a fluid battle and how to fight it in what they called "depth," both geographically and in time. For the first problem, as advanced principally by Tukachevsky, the state must essentially prepare for total war through a total and ongoing mobilization. As will be argued later in this chapter, the French has established a nascent version of what James Schneider has called "the warfare state," although Schneider argues that its first mature or perfect form was that imagined by the interwar Soviet operational art theorists and then implemented in the war against invading Nazi armies.[21]

The idea of "deep battle" also permeates the Soviet writing on operational art. The two most prolific authors on this score were V. K. Triandafillov and G. S. Isserson.[22] Of the two, Isserson is perhaps more important for our purposes. Both of these Soviet officers believed that modern operations in the future total war must be executed throughout the depth of the battlefield. At the operational level, this meant that one must plan for both simultaneous and successive operations, to meet the enemy's tactical, operational, and strategic reserves. These reserves, or defensive "echelons" as Isserson called them, were to be met and dealt with in turn by offensive operational echelons. Isserson was perhaps most explicit in laying a formula for how to do this:

> A modern operation essentially elicits *distributed* efforts in time, thereby conditioning strategy. . . . Modern operational *echelonment* is the sequential and continuous increase of operational efforts aimed at breaking enemy resistance through its whole *depth*. . . . While deploying for a modern deep operation, it is necessary to calculate forces and means both along the linear dimension of a front and in the new dimension of depth.[23] [emphasis added]

The upshot of the analysis to be used in this study, then, will be that the criteria for examining the elements of operational art in the great Napoleonic campaigns will consist of looking at the depth of operational engagements and battles, use of "warfare state" methods such as the mobilization of the entire society, and the presence of multiple operational and even strategic echelons during these campaigns. Isserson also used the term *distributed* in describing modern operations, and the next section provides a brief discussion of how that concept will be used in this analysis.

SCHNEIDER'S FRAMEWORK FOR OPERATIONAL ART

James Schneider has argued that the emergence of operational art is tied closely to the emergence of technologies in the 19th century that enhanced movement, firepower, and communications—the steam engine, mass-produced rifles and high explosives, and the telegraph. These technologies certainly influenced the tempo and complexity of modern war, but they did not foreclose a more nuanced view of operational art, that is, the conduct and practice of war at the operational level by military men, usually officers of field grade (major) or higher. Accordingly, we shall examine Schneider's model for assessing operational art without tying it too closely to 19th-century technological developments.

Schneider has made the assertion that it was during the American Civil War that "a new style of warfare emerged," which he characterizes as operational art. Schneider presents eight "key attributes" that allow for the "fullest expression" of operational art. He identifies these attributes as a distributed operation, distributed campaign, continuous logistics, instantaneous command and control, operationally durable formations, operational vision, a distributed enemy, and distributed deployment.[24]

Before a discussion of the course of these campaigns, we must present the definitions of Schneider's eight "attributes" or elements of operational art. This will allow the reader to identify the appearance of any or part of these elements as we review the history. His first element is the *distributed operation*. "Operational art is characterized first by the distributed operation: an ensemble of deep maneuvers and distributed battles extended in space and time but unified by a common aim." This definition includes the existence of "temporal and spatial distribution" in the conduct of a campaign. The relationship between this "characteristic" and what the U.S. Army labels *deep operations* is in fact the fundamental discriminator between operational art and classical strategy.[25] Modern U.S. Army doctrine defines deep operations as those that "engage enemy forces throughout the depth of the battle area and achieve decisive results rapidly."[26] Therefore, the existence of "deep operations" will be a key indicator that an operation is distributed.

Schneider's next element is the *distributed campaign*, which he characterizes as the distributed operation's "final structure," the product of the operational artist's vision. Several characteristics of distributed campaigns include the subordination of battle to maneuver, and then, "Because modern warfare emphasizes battles and maneuver, distributed campaigns are inherently exhaustive."[27] *Continuous logistics* is a third element found in operational art. "In order for a modern industrial army in a theater of operations to maintain a militarily effective presence, its logistics system must be continuous." Schneider makes the analogy that in modern warfare the laws of Newtonian physics are replaced by the laws of fluid dynamics. The dynamics of war using the operational art construct more closely resemble fluid laws governing distributed operations and modern industrial warfare just as Newton and classical physics (mass, accelerations, force) were superseded by the laws of fluid dynamics and later the laws of probability and quantum mechanics.[28]

The fourth element is *instantaneous command and control.* Distributed operations generate more information; therefore, command and control must be instantaneous. Schneider's criterion here relies on the technology of the telegraph in order for operational art to be in its most mature form. In this discussion, Schneider uses the battles of Jena-Auerstadt (1806) and Waterloo (1815) as examples to prove his point. Schneider hypothesizes that had Napoleon had telegraph at either of these two engagements the results may have been very different. In fact, a very good argument can be made on both counts that telegraphic communication would not have affected the outcome of either battle. Another point to make is that a form of command and control (C2) that the U.S. Army calls "mission command" obviates the need for constant, instantaneous communications via wire or wireless means on today's battlefield and requires no technology at all, but rather clear, mission-focused orders in order to give subordinates the ability to exercise disciplined initiative in the absence of such communication.[29] A better measure might be *the effectiveness* of the information transfer, vice its speed of transmission. Schneider's own example of the Battle of Chancellorsville supports this proposition. Even with telegraph, the command-and-control climate and its effectiveness, not the speed of information transfer, helped make the difference at Chancellorsville. It is not the speed with which a commander communicates, but the *understanding* of his intent by his subordinates and their ability to execute that intent in the presence of an enemy.

The fifth element that Schneider uses is *operationally durable formations.* These are "formations capable of conducting indefinitely a succession of distributed operations."[30] According to Schneider, these formations were first seen during the Civil War as field armies. Russell Weigley, on the other hand, hints that this type of force, one that relied on attrition and exhaustion, had already been developing for some time. The Russians may have had operationally durable formations as early as the Seven Years' War (1756–1763)—the time when European armies were becoming difficult, if not impossible, to destroy in battle. This dynamic will be another key component in this analysis, since many scholars focus precisely on the Napoleonic period as giving birth to such formations as a result of the total mobilization of the nation-state.

Until 1809, the French armies almost always had an asymmetric advantage over their opponents because their armies were durable as a result of mass conscript, their patriotism to serve the colors, and Napoleon's institutionalization of the new combined arms organizations—the division and army corps—which combined artillery, infantry, and cavalry to make these units even more durable should they encounter an enemy unit on their own. Although Napoleon's opponents adopted these measures—in 1809, the first such anti-French army was put in the field by Austria—it was rare that a Napoleonic Field Army was destroyed in one day's battle or even in one campaign.[31] Sir Charles Oman's monumental work on the Peninsular War in Spain (1808–1814) emphasizes Duke of Wellington's astonishment regarding the durability of French armies. The Duke of Wellington freely admitted that he never saw anything like the recovery of the French armies after their retreat from Portugal after their defeat at Salamanca, and Marshal Soult's unexpected offensive across the Pyrenees with a French army that had been shattered at Vitoria some months earlier.[32] Another of Schneider's criteria for judging the maturity of the element of operationally durable formations is "the emergence of the army group."[33] As we shall see, the Napoleonic era provided abundant examples of this type of formation, especially after 1805.

Operational vision, a *distributed enemy*, and *distributed deployment* comprise the final three elements, or criteria, that Schneider uses to characterize operational art. Schneider associates operational vision "with mental agility, the ability to react to incoming information faster than it arrives." Schneider adds that a staff *plays* a "crucial role . . . in sustaining that vision." Therefore, this book uses the following definition for operational vision: seeing beyond local operations and results to a future end-state that accomplishes strategic goals. This vision must be exercised in a *distributed* environment, which is often a very broad front with some depth, similar to that described by Schneider, in order for it to be operational.

As discussed earlier under instantaneous command and control, Schneider indirectly gave Napoleon credit for being a *distributed enemy*, by using two of his campaigns as examples relevant to a discussion on instantaneous command and control. However, for mature operational art to present we must also look for instances where both sides operated in a distributed manner, not just one side or the other. In almost every instance where French armies fought a counterinsurgency (COIN in today's parlance) against guerillas and irregulars—in the Alps, in Spain and Portugal, in southern Italy, in Russia in 1812, and in Germany in 1813—their enemies were distributed. However, because the French also fought against conventional forces in all but one of these COIN campaigns—a type of conflict known as compound warfare—this aspect of operational art will be discussed as a component of these campaigns rather than separately from them.[34]

Schneider provides no precise definition for *distributed deployment*, but does tie this element to the industrial revolution and a nation's capacity to wage war. Nations must "defend—and seize—the resource and production base." This element would seem to be a given if the operation is distributed and nations have undergone the kind of total mobilization we see occurring in the latter half of the Napoleonic Wars. But its earliest expression may have been the announcement of the levee en mass and Lazare Carnot's efforts in 1793–1794. At any rate, we will define distributed deployment as deployment in breadth and depth, both military and economic, in support of operations of war. This element can also include a maritime component, as demonstrated most effectively by Great Britain during these wars. Because of the Royal Navy's reflection of this, we see then that operational art is not only limited to land but can also encompass purely naval as well as combined (joint) operations.[35]

* * *

Invention, whether for abstract scientific questions or for humdrum day-to-day practical issues, often springs from the solution of a specific, seemingly intractable problem or set of problems. Usually there is no need to invent a new approach if the problem can be solved with existing means. Such was the case with the emergence of what I have been referring to as operational art. The thinking and writing about operational art, like the calculus, emerged from several different contexts in separate parts of the world during a relatively short time span of history. Just as Gottfried Leibniz and Issac Newton, among others, sought to solve the seemingly intractable problem of figuring out complex kinematic trajectories and the area under a curve, so too did the operational theorists labor to solve what they saw as emergent problems related to the complexity of modern, industrial war.[36]

An operational examination of Napoleonic campaigns has value because so many of their characteristics resemble current American military thought and practice. The Napoleonic period resembles a veritable laboratory of coalition and operational-level warfare and provides a means of understanding the lessons of a historic period as a way to help understanding the dynamics of coalitions and operational art. Coalition warfare, because of its breadth and depth of complexity and often its scope of military activity, may in fact be a context that highlights the practice of operational art. Additionally, the primacy of politics, maneuver, mission command, and early forms of deep operations emphasize areas of similar importance in current U.S. Army doctrine today. Although the solutions and operations cannot be applied directly to today's challenges, the process of operational art, and the dynamics of human behavior on the grand scale of history, can be understood so as to better understand modern challenges and the process that could lead to their resolution.

In summary, the problem of modern operations is not new. Since the beginning of the modern era, roughly in the 16th century, the complexity of war and expansion of its scope have constantly challenged those who think, write, and practice it to try to keep up with its evolving nature. In the pages ahead, the fundamental goal will be to describe these campaigns, armies, and leaders, using the lens of operational art. Descriptive details of battles, leaders, and equipment will be economized in order to more fully look at operational structures, maneuvers, and consequences. No model, or framework, can completely describe any past reality, but in using new ones, some new insights might be gained.

TWO

THE WAR OF THE FIRST COALITION AND THE BIRTH OF MODERN OPERATIONAL ART

CHRONOLOGY

Aug 15, 1769	Napoleone di Buonaparte [as christened] was born in Corsica.
1774	Louis XVI crowned king of France.
1785	Napoleon was graduated from military school in Paris and commissioned as an artillery sublieutenant.
Jun 1789	Estates General, convened by the king for the first time since 1614 because of the government's financial crisis, declared itself a national assembly. A series of laws and pronouncements issued during the following weeks started France down the road toward democratic rule.
Jul 14, 1789	Storming of the Bastille.
Jun 1791	Louis XVI was arrested as he attempted to flee Paris.
Apr 1792	France declared war on Austria. Austria and Prussia invaded France. War of the First Coalition began.
Sep 1792	French forces defeated the Prussians at Valmy and saved the Revolution. French forces subsequently invaded the Rhineland, Savoy (Italy), and the Austrian Netherlands (Belgium).
Nov 1792	Dumouriez defeated Austrians at Jemappes.
Jan 1793	King Louis was executed. Great Britain and the United Provinces (the Netherlands) declared war on France.

Mar 1793	Spain entered war against France as do Sardinia and Naples and several other states. Dumouriez lost at Neerwinden against Austrians and abandoned Belgium.
1793	A serious Royalist-backed revolt broke out in the Vendee region. France faced a civil war as well.
Aug 1793	Committee of Public Safety declared the *levée en masse* to defend *La Patrie* (the fatherland). French armies swell to almost a million men under arms.
1793–1794	Reign of revolutionary terror; French populace mobilized to fight a coalition of European monarchies and rebels in the Vendee.
Jul 1794	Thermidorian coup ended the Terror and restored a more moderate government.
1794	Napoleon Bonaparte was promoted to general de brigade (brigadier general). Holy Roman Empire declared war on France, bringing in smaller German states to join the First Coalition.
1795	Prussia and Spain both withdrew from the war after defeats at Fleurus and San Sebastián, respectively.
Oct 1795	Napoleon Bonaparte was promoted to *general de division* (major general).
1796	Bonaparte assumed command of the French army of Italy.
Oct 1796	Spain declared war on Great Britain.
Apr 1796	Bonaparte attacked Austrians and Piedmontese, winning the battles of Montenotte, Dego, and Mondovi in 10 days.
Apr 28	King of Savoy (Piedmont and Sardinia) signed initial peace with Bonaparte.
May 10	Battle at the Bridge of Lodi—Bonaparte chased Austrians across the Adda River and occupied Milan.
May–Jun	After forcing the line of the Mincio River, Bonaparte attempted to seize the fortress of Mantua by a coup de main. When this failed, he settled in for a siege that eventually lasted eight months off and on.
Jul 31	Bonaparte lifted the siege of Mantua.
Aug 3–5	Bonaparte's army defeated Austrians at Lonato and Castiglione.
Mid-Aug	Würmser retreated to Trent after revictualing Mantua. French siege resumed.
Sep 8	Battle of Bassano, French victory.
Sep 12	Würmser escaped into Mantua, which remained under siege.
Jul–Oct	French armies under Moreau and Jourdan are forced back over the Rhine by Archduke Charles after an initial offensive. Austrian reserves were freed up to assist against Bonaparte in Italy.
Nov 15–17	Battle of Arcola, Bonaparte defeated Alvintzy.
Jan 14–15, 1797	Battle of Rivoli, Bonaparte decisively defeated Alvintzy.
Feb 2	Mantua capitulated.
Mar	Bonaparte resumed offensive against Archduke Charles.
Apr 18	Bonaparte forced Austrians to sign an initial peace at Leoben.

In 1792, the French Revolution developed into a general European war known variously as the Wars of the French Revolution or alternatively as the War of the First Coalition (against France). It thus encompasses the broad canvas of campaigns as France fielded Europe's first modern "armies of the people" under the leadership of men who rose under the system known as "careers open to talent" in France. It concluded with the key campaign by the young General Napoleon Bonaparte in northern Italy that brought about the first general truce, excluding Great Britain, and an end to the War of the First Coalition.[1]

Napoleon Bonaparte consolidated and further institutionalized the reforms that gave birth to the type of warfare and the wars that came to bear his name. A better name for these wars, as already alluded to, is the French Revolutionary *and* Napoleonic Wars. Because of this, students of this period of military history often become confused about what Napoleon brings to the table as a military innovator. Napoleon implied that he did not end the Revolution; rather he tamed its excesses and institutionalized what was good in it (laws, education, religious tolerance, egalitarianism, etc.).[2] Similarly, he did not create a military revolution; rather he further consolidated, refined, and improved upon the existing system as a means to further what he saw as his own progressive policies. Military staffs, combined arms divisions, the *levée en masse*, new tactics, excellent artillery, and a host of other reforms and innovations that are often associated with Napoleon in fact predated him.[3] In short, he inherited with his first army-level command an excellent system that had already been in place for some time and then improved upon it. In fact, the young General Bonaparte did not implement many of his military innovations (mostly organizational and administrative), as we shall see, until *after* his first several campaigns. There was an existing way of doing business and he very much used the system in place to achieve his first successes. Put another way, Napoleon did not achieve success because he brought much that was new to the battlefield beyond his own strength of purpose and competent generalship. The military tools he inherited from the old regime of the Bourbons and those created by the conditions of the Revolution already had within them many of the operational building blocks that soon became the components, metaphorically the paintbrushes and oils, for his operational art. It is to those components this discussion turns before looking at Napoleon Bonaparte's first operational masterpiece in Italy and Austria in 1796–1797.

THE OPERATIONAL TOOLS OF THE TRADE PRIOR 1796

Defeat drives reform, and failure often leads to innovation. Historian Steven Ross put it best when he wrote of the French defeat by the British and the Prussians in the middle of the 18th Century: "For France the Seven Years' War was an unmitigated disaster. The army entered the war without enthusiasm, fought without distinction, and emerged without victory."[4] From this cataclysm flowed the impetus for reforms that led to the tools that the revolutionaries of 1792 would take up in their war against the rest of Europe.

The first innovation to flow from these reforms was the creation of the *division* as a standing, and not ad hoc, military formation. Originally the French had used the term to specify independent formations from the main army, but it came to mean the administrative division into which France was divided for military purposes. It was not long, however, until a separate system of combat divisions, aligned with the administrative divisions, came into being. These divisions were composed of the infantry forces of the administrative district with clear military responsibilities for defense. The subunits of organization under these "divisions" were regiments composed of battalions, but the battalion was the basic infantry building block and included anywhere between 500 and 1,000 rank and file. The reforms of the Comte de St. Germain (Count of St. Germain) in 1776 had established two battalions per regiment, usually one for field service and the other a cadre, or depot, battalion.[5]

As mentioned, these divisions were normally not what we would call combined, or mixed, arms formations. Sometimes there was attached cavalry but rarely did they include engineers (*Genie*) or artillery organic (inclusive) under the oversight and control of the divisional commanders, whose duties were mostly administrative in nature. Interestingly, divisional commanders were encouraged to train with other arms to improve coordination on the battlefield, one of the many lessons learned the hard way on the battlefields of North America and Europe.[6]

If one wants to see what a postwar but prerevolution French combat division might have looked like in the field, with various arms attached, all one needs to do is examine the organization of the Comte de Rochambeau's component of the Anglo-American army during the Yorktown Campaign. Rochambeau's force numbered approximately 7,800 troops and included infantry, cavalry, engineers (to help with the siege), and abundant artillery. Bonaparte's future chief of staff, the famous Marshal Louis-Alexandre Berthier, served in this formation as an engineer officer.[7]

Berthier's prerevolutionary service highlights several points about the French army of this period. First, one could be an officer and not be from the aristocracy if he served in the "scientific" arms such as the artillery or the engineers. Second, Berthier was from the middle class, his father also being an engineer officer, and so foreshadowed a trend that the Revolution would formalize using the sons of the emerging middle class as officers. A Counter-Reformation of sorts would eventually occur in the last years of the French dynasty against Comte de St. Germain's reforms, which resulted in even these officers being required to have an aristocratic pedigree, but the point to be made is that for these "disesteemed" branches, the French army already had a system of merit in place if the officer was content to never achieve high command or rank or practice his trade outside his selected "branch" of service.[8] The "career open to talents" existed in embryonic form prior to the Revolution.

Finally, Berthier's status actually changed prior to the revolution, when he became part of a new branch in the French officer corps conceived by another prerevolutionary reformer, Pierre-Joseph Bourcet. As part of the reforms ongoing in the French army, Bourcet's reforms were adopted after his death in 1780 by the French minister of war Philippe Henri de Ségur. Ségur created the *Corps d'Etat Major* (where the modern rank of major comes from). "The officers chosen for it were to be trained in topography, history, geography, preparation of unit histories, reconnaissance, and the science of the art of war."[9] Thus was instituted the concept of the modern staff officer as well as the requirement to come up with a system to educate these men. They became known as "topos" in the Royal Army, and Berthier was among those chosen to become one. With the revolution, this promising organization, which was regarded as a bastion of aristocracy, was abolished, but was reestablished in 1790 under another name. By 1792, each army had a functioning staff with a chief of staff and half a dozen *adjutants-generaux* (staff adjutants). This staff could be augmented by attaching officers from subordinate units as required (e.g., engineers, cavalry, and artillery "aides").[10]

A year prior to the fiscal crisis that led to the French Revolution, more organizational reforms were implemented under Louis XVI. First, the brigade was established as a permanent formation with permanently assigned regiments. Additionally, these brigades "were formed into 21 combat divisions." It has been speculated that these organizational innovations may not have lasted had the French Revolution not occurred and served as a forcing function to not only leave them in place but also to expand on them as the paradigm changed from one of cautious conservative innovation to sometimes incautious Revolutionary experimentation. However, these divisions were still almost entirely infantry formations, but they did occasionally train in the field with the other units.[11]

Two other areas that saw reform had to do with light infantry tactics and artillery. By the time of the Revolution, France possessed the most rationalized and well-organized artillery in Europe due to the famous "Gribeauval" system that reduced the French guns into four standard calibers and made the actual equipment lighter and more maneuverable. Additionally, an officer named Jean du Teil, part of a group of increasingly professional and intellectual French officers, advocated using these more mobile guns primarily against infantry, not the enemy's guns. With the artillery we see also the French officer education system at its best, requiring actual merit by its officers, such that young cadets with less than stellar pedigrees like a talented young Corsican named Napoleone di Buonaparte at Brienne, later at the *L'Ecole Militaire* in Paris, and finally at the Artillery School in Auxonne where he met the great du Teil in person, could thrive academically whereas the blockheads and idlers fell by the wayside.[12]

The other area of reform had to do with light infantry tactics, an area that the French had been poor at in the Seven Years' War and that the American Revolutionary experience seemed to indicate was of more importance. The epitome of tactics had come to be represented by the linear Prussian system of Frederick the Great involving strict discipline and the elimination of initiative from all but the most senior officers on a battlefield. Among the officers discussing these reforms were du Teil, the Comte de Guibert, and the Duc de Broglie, men of the Enlightenment who believed that France could excel in military affairs as she did in all other areas of intellectual activity (at least from a French perspective!) De Broglie, along with the Duc de Choiseul, had been instrumental in the reforms leading to the establishment of the division. However, it was Guibert who foreshadowed the citizen-soldier army with his "General Essay on Tactics," and who identified the value of the division as a means to advance in several columns on several routes instead of the standard route of march that placed a single army along a single road. Guibert died before the Revolutionary Wars came, but his ideas were in circulation—despite his having disowned the most innovative ones several years later in a second treatise.[13] The key point to remember, though, is that, with the exception of the American Revolution, the old regime in France never got a chance to try any of its new tools out in a large-scale fight in Europe. It fell to the military of the new French Republic—essentially the same army we have just described without the bulk of its aristocratic officer corps—to employ these tools in battle.

As can be seen, the focus here has been almost entirely on the French; however, other states (not yet nations perhaps) were also developing and innovating, trying to come to grips with what the future of war might hold. The antithesis of the French, at least in most of the writing on the topic, seems to be the military, which was regarded as the best of its day and which all others seemed to try to emulate— the army of Prussia created by Frederick the Great and his father Frederick William II. The stereotype of this army is one of draconian discipline, rigid linear tactics, and an extreme limitation on initiative by the commanding general, the acknowledged military genius of *Der König* (king), Frederick the Great. In actuality, the other armies of the period were also attempting to come to grips with reform and innovation, trying to maintain a perceived advantage (Prussia and Russia) or to correct for recent setbacks (Great Britain, and to some degree Austria). Russia, and less so Britain and Austria would bring some originality and creativity to the tactical and even operational side of war. Russia, in particular, with its solid record of success against Frederick in the Seven Years' War, brought much to the table, as we shall see in chapter 3, when she entered the lists against the new and reanimated French juggernaut. The Prussians seemed to believe that in the wake of the great King's death, they, too, might have to make some adjustments, albeit minor.

Looking ahead, none of the other great powers seemed to match France after the wars began in light of the institutional impact of their own modest reforms. The results from the campaign trails certainly seemed to bear this out, especially after Bonaparte emerged on the scene, but other French generals would be quite successful with these tools, too, such as Masséna and Moreau. The reforms of the opponents of France, then, will be discussed as these various adversaries arrive on the stage. Only France employed divisions as permanent organizational units in combat, but the division was all infantry and not yet a combined arms organization, it had no integral cavalry, artillery, or engineers. French artillery was perhaps the best in Europe, but was yet untested on the Revolutionary battlefield. In sum, French military reform had yet to be tested in the heat of battle.

FORGED BY FIRE: THE EVOLUTION OF OPERATIONAL ART IN THE EARLY YEARS OF THE REVOLUTION

Although the Revolution began in 1789, it was only in 1792 that the French Republic lurched ill-advisedly into war against the other European powers. The increasingly factionalized polity of France (it was not really a unified government) welcomed the war as a means to help solve pressing domestic politics while others inside the government, such as the foreign minister Charles-François Dumouriez, hoped a quick victory might help position them and their party as the "power behind the throne" in the new Republic (the King had not yet been executed). Accordingly, the government pressured the King into asking the National Assembly to declare war on Austria in April 1792. Austria reciprocated and her ally Prussia joined her against France in what became known to history as the First Coalition. It eventually included virtually every other great power in Europe, with the exception of Russia, which was busy partitioning Poland with Austria and Prussia during this period and then pacifying her new Polish subjects.[14]

War has a way of increasing a nation's learning curve. Unfortunately, for France's opponents, the French military seemed to learn more quickly than they did—perhaps this was because she was unencumbered by the conservative institutional constraints of the old regime. However, if one were to go to the 18th-century version of Las Vegas and make a bet based on the numbers, France would surely have been the long shot of long shots in 1792 to win against two of the foremost military powers of Europe—Austria and Prussia.

At the time the army was asked to defend the Republic, it had been seriously weakened by the Revolution. Especially hard hit was the officer corps, although discipline among the enlisted ranks had suffered, too. By 1792, over 30,000 troops had deserted and the officer corps had been halved from its pre-Revolutionary strength of

9,000. By 1794, more than two-thirds of the officers from the old army were gone, having emigrated, been executed, or imprisoned. These numbers were even higher in the cavalry and infantry, where only aristocrats had been allowed to serve. In the new National Guard, created as a militia to protect the Revolution, and the volunteer units such as the *Hussars de la Liberté*, the officers were elected and there was little discipline other than enthusiasm and a desire for *élan*. Relations between the National Guard, the newly called-up volunteer units, and the old Royalist cadres were poor, none trusting the other. No one, other than the French and opportunists such as Dumouriez, considered that these forces might be able to stand up to the professional armies of Austria and Prussia in 1792.[15]

On the positive side, the army of long-term professionals that remained was, for the most part, committed to the Revolution. This was especially true of the officers, although the enlisted troops who remained were committed as well now that new opportunities never imagined had opened up. Many of the remaining Royal Army noncommissioned officers would later rise to the highest levels of the French military establishment as marshals and generals, leading to the truism that every French *soldat* had a potential marshal's baton in his knapsack—another indicator of the impact of the egalitarianism and the "career open to talents."[16] Nonetheless, aristocratic officers of the old regime commanded this army in the first battles that it fought and these officers fought in the same style as they had fought for most of the century, with the infantry in lines, cavalry in reserve or on the wings, and the artillery centrally located and controlled. At Valmy on September 20, 1792, the great victory just east of Paris that saved the Revolution, the commander, General François Kellerman, fought using the prewar infantry brigades, not divisions, with no integrated artillery. Kellerman famously repelled the Prussian attack in foggy weather with his excellently served cannon.[17] This battle, however, did not end the war, but simply ended the campaign and staved off disaster.

The period after Valmy briefly saw France on the offensive against its foes as the armies swelled with volunteers eager for glory. A defeat of the Austrians later that fall delivered the Austrian Netherlands (Belgium) to Dumouriez and the conquering French. However, French success and conquests brought more nations into the fray against them. After the execution of Louis XVI in January 1793, Great Britain, the United Provinces (the Dutch), and later the Kingdoms of Spain, Naples, and Sardinia all entered the war against France. Additionally, the French were now faced by a skilled Austrian commander of the old school, the Prince of Coburg. As Dumouriez invaded the Netherlands, Coburg struck at his flanks, forcing him to pull back and then decisively defeated him at the battle of Neerwinden in Belgium on March 18, 1793. The French armies were expelled from their recent conquests and Dumouriez defected to the anti-French coalition when his army refused to march on Paris and overthrow the government.[18]

The importance of these events for our purposes is that they created a crisis for the government, resulting in the call up of 300,000 conscripts during the month of the military disaster (March 1793). By August, the military crisis was so serious that the new Committee for Public Safety issued the famous *levée en masse* that created Europe's first mass conscript "army of the people." The government soon had 14 field armies with over 660,000 men in the ranks, and by 1794, it would have over a million armed troops. The rapid expansion of the army to protect *La Patrie* (the fatherland) combined with the shortage of officers to drive organizational innovation. Necessity, as usual, became the mother of invention. With fewer officers and more rankers, the French needed ways to both integrate the new untrained levies with the veteran Royalist troops, as well as control the much larger armies that were created to defend the Revolution.[19] The span of control for these armies organized into brigades was too great because the armies were so large. Fortunately, the French had the example of the prewar division ready to hand to implement and employ.

Much experimentation took place, but eventually the regiment was replaced for a time by something known as the demi-brigade, a formation that married up two of the new conscript or volunteer battalions with one of the old royalist battalions, giving a total strength of three and sometimes four battalions in a demi-brigade. Two of these demi-brigades were in turn supposed to be combined into a larger full brigade. Because of the shortage of officers to command brigades, this did not happen and often as not the demi-brigades were grouped together into divisions of varying size under one general, instead of having the intermediate level of the brigade requiring additional generals between the demi-brigade and division levels. These reforms took place under Lazare Carnot, a former engineer captain turned politician, in 1794. The revolutionary government created the new rank of general of division. These generals were then given their own artillery and often a regiment of cavalry, and with this the true combined arms division was born. Later, engineers could be added to the complement of the division depending on its mission. Sizes for the divisions often varied, but they usually consisted of at least three demi-brigades with supporting artillery. They usually ran anywhere between 4 and 10,000 men in strength and the size tended to directly reflect the army commander's opinion or confidence in his divisional commander.[20]

While the demi-brigades became permanent administrative entities, each with its own number and with depot battalions assigned to French Military Districts (the former administrative divisions), the divisions remained formations that were often formed in the field and then assigned to the generals at hand, who sometimes only had the rank of general of brigade (again due to the shortage of officers). This dynamic explains why officers like Bonaparte went from being captains to generals in two to three years, or even less. Also, the officers used to command at these new levels could come from any branch, not just the infantry or cavalry. Again, the artillery officer Napoleon Bonaparte is prima facie evidence of the best sort to highlight the opening of high command to officers from all parts of the army. Later, near the end of the Wars of the Revolution, the divisions became somewhat more permanent, taking the name of their commanding general (e.g., Masséna's Division in the Army of Italy). Cavalry divisions also became combined arms formations as horse artillery was placed, again, under the direct control of the cavalry division commander. By January of 1796, on the eve of Bonaparte's Italian campaign, there were over 140 demi-brigades. Theoretically, these could be used to generate something on the order of 40 divisions to populate over a dozen French field armies. From 1794 on, the combined arms division was well established and served as the primary unit of maneuver, both operationally and on the battlefield.[21]

By 1796, most French generals were well versed in commanding divisions. At the level of field army commander, the most senior generals learned in an evolutionary manner how to get the best out of these units in campaign after campaign. As discussed, divisions could march on separate routes, which increased their flexibility of employment and eased logistical arrangements, opening up broader areas to forage to ease the always poor French logistical arrangement characteristic of the period. They were large enough that in a pinch they could defend themselves or conduct a fighting retreat should they encounter enemy formations as large or larger than themselves, often for several hours if not longer. Because of this flexibility and *durability*, commanders learned how to fight encounter battles and meeting engagements, advancing on broader fronts than in previous times and then feeding the other divisions into the fight in an echeloned manner, often on the flanks, as later formalized for higher echelons (corps and armies) by the Soviet operational art theorists.[22]

Another area that the creation of these mass French armies stimulated was the size and number of staffs in the French field armies. The creation of the division led to the establishment of staffs on the model institutionalized in 1792 for the army commanders. Specialized staff officers, if available, were appointed as chiefs of staff. By 1799, the number of these specialist staff officers, the "topos" of the old regime, increased to 110. Also, not all of the assigned officers to these new staffs were formally trained given the shortage of officers; many of them learned "on the job" under the veteran chiefs of staff like Berthier. Often the only requirement was literacy and even a rough-and-ready former cavalry trooper like Ney was used as a staff officer because of his ability to speak and write German as well as French, although he would later take on the young Jomini as his chief of staff to do this sort of necessary, but often tedious work.[23] All of these developments, plus the constant campaigning, had the result of creating a very large body of young, experienced, and toughened officers who would form under Bonaparte an extremely valuable cadre of *professional* officers to conquer, rule, and administer an Empire.

The more numerous armies and their often large size of around 50,000 to 70,000 troops each resulted in the achievement of two criteria we have already discussed that were needed for operational art to exist—*operationally durable armies* and *distributed deployment* (see chapter 1). With over a dozen armies the French could protect key areas of the nation that supported the war effort in-depth with operational reserves of army size to the rear or in a neighboring theater. This constituted an embryonic system of distributed deployment, before Napoleon Bonaparte even came on the scene.

The period from 1794 to 1796, although characterized by a trend of increasing French victories, nevertheless resulted in a strategic stalemate in both Italy and Germany, where the Austrians and their few German and Italian allies remained in the field against the often poorly supported French armies, which no longer defended France, but challenged the enemy in their own territories of influence.[24] French armies lost battles as much as they won them, but warfare had reached a point where these French armies could not be shattered or destroyed in one day or even one campaign. When the French won, they often found that they lacked sufficient combat power to destroy their enemies and achieve decision. The French commanders had push them back and then they culminated logistically, always a problem for the ill-supplied French armies. Another reason the French had difficulty exploiting their battlefield victories was their generally poor cavalry, which often found itself bested by the excellent Austrian or Prussian cavalry when it tried to pursue the beaten allied infantry. The armies would then reconstitute both logistically and in terms of replacing manpower lost or sick due to the campaign. However, because of conscription and the fact that French soldiers might leave the ranks but usually returned to the colors for the march and the battle, the French seemed always able to reconstitute their armies for another campaign season quicker than their opponents. The Duke of Wellington, in particular, would be bedeviled by this characteristic during his campaigns in Spain.[25]

EARLY OPERATIONAL ART: NAPOLEON IN ITALY

The First Coalition was an 18th-century response to the catharsis that defined 19th-century Europe—the French Revolution. The apparent goal of its protagonists—the overthrow of the revolutionary French government—was never shared by all the participants simultaneously. Implicit in this goal was the restoration of the territorial status quo and a Bourbon monarchy. The major powers eventually opposed to France—Spain, Austria, Prussia, and Great Britain—were more interested in the enhancement of their territorial integrity or position in Europe (or Empire in Britain's case) than in the restoration of the House of Bourbon to the throne of France. No unity of command or synchronization of the disparate military efforts going on in the various theaters from the Pyrenees to Germany to Flanders existed. As a result, each nation pursued its goals for all practical purposes without regard to the others. Despite some local victories, the First Coalition slowly fell apart as the huge French armies generated by the *levée en masse* defeated or stalemated each nation in turn and then began to operate uniformly in the enemy's territory or sphere influence—principally Germany and Italy by 1796.

The constant wars since 1792 also led to war weariness in France and the beginning of the phenomenon of "draft dodging," so much so that French victories on the Spanish, Rhine, and Italian fronts in 1795 led the French government to compromise more readily than its predecessors. Therefore, in 1795, Prussia, Spain, and the Grand Duchy of Tuscany signed peace treaties with France and left the war.[26] Now, only the most implacable opponents, Great Britain and Austria, remained. It is to the Austrian opponent that we now turn and thus the young Napoleon Bonaparte's campaign, which destroyed Austrian military resistance and brought an end to the First Coalition in 1797.[27]

When Napoleon Bonaparte was finally given the command he sought in early 1796, it was of the most ragged and ill-supported French army—the Army of Italy. The army had been neglected for much of 1795, was ill-fed, and unpaid. The new "political" general was only 26 years old and was elevated over some of the most veteran and competent generals in the French army—Andre Masséna, Charles-Pierre Augereau, Jean Serurier, and A.E.F. La Harpe. These men were to command Bonaparte's divisions with his army's cavalry under the veteran Baron Stendhal. Additionally, he had Berthier, possibly the best staff officer alive, as his chief of staff. Historians have frequently commented upon Bonaparte's fortune in inheriting such a talented group of subordinates, but initially they took what might be regarded as a "wait-and-see" attitude toward the young upstart from Paris. It was not long, though, before they gave him their respect and their best efforts. Bonaparte brought with him many young officers of his own choosing that later gained great fame—Andoche Junot, Auguste-Frederic Marmont, and Joachim Murat being among his closest associates and friends. He also famously included a trunk of books by the likes of Caesar, Plutarch, and Xenophon and other military histories among his personal belongings for the campaign trail—for he was a voracious reader and slept little.[28]

Napoleon Bonaparte's army was not the main effort for the French strategy in 1796. The brainchild of Lazare Carnot, this strategy's goal was to conduct simultaneous offensives in both Italy and Germany that, like the arms of a great pincer, would close upon the Austrian armies somewhere in the vicinity of the Austrian Tyrol. The goal for each offensive was to tie down its Austrian counterpart so that the Austrians could reinforce neither theater to support the other. Given the distances between the two theaters, and the fact that the impassable Alps also lay between them, it was a rather far-fetched, ambitious scheme. The bulk of the French military power would be north of the Danube River in Germany in two armies commanded by Generals Moreau and Jourdan. Neither of these two generals had authority over the other and both answered directly to Paris.[29] However, from an operational standpoint, the idea of a coordinated offensive on such a broad scope using the bulk of four armies (a fourth smaller army was to connect the operations with each other in Switzerland), did give the entire concept the characteristic of a distributed deployment. Logistics would almost certainly be a weakness and the idea of instantaneous command and control is laughable—except for the fact that Bonaparte would operate almost entirely without regard to the other armies, following his own vision of how best to achieve victory. His goal was first to defeat the Austrians in his theater, and if the armies in Germany failed, he might find that he could turn his theater into the main effort of the 1796 campaign. If in fact the latter outcome was his goal, he was to be amply rewarded. This in turn might lead to his making as being among the first rank of French generals—not just as a political favorite of the regime.

When one looks back on the campaign one finds that it can be divided into three phases: an initial offensive phase in Savoy and Lombardy, a lengthy phase with Bonaparte on the defensive as he conducted the siege of Mantua, and finally a closing phase with Bonaparte on the offensive again into Austrian territory. This kind of operational flow matches much more with the idea of the lengthy campaign of attrition and exhaustion preached by Delbrück as one means to strategy, and by Svechin as well. However, some historians, including Jomini, have instead labeled it as *model of maneuver and decisive battle*. Few of the battles were decisive, they were often engagements, lasting several days in some instances, and had the result of leading to more shifting and maneuvering. They were often repulses and rebuffs that gained time as Bonaparte ground down the Austrian army, which gamely returned to the fray again and again until it was finally administered the *coup de grâce* at Rivoli, putting it permanently on the defensive. Like Frederick the Great's later campaigns, Bonaparte maneuvered against his often superior enemies, taking full advantage in several cases of the superiority of the operational and tactical defense as he defended gains that the Austrians felt they had to reclaim.

Bonaparte's opponents were not as antiquated and overmatched as has often been portrayed. Certainly the Piedmontese, led by the Austrian general (on loan) Michael Freiherr von Colli-Marchini, did not have their heart in any sort of warfare requiring much of them, and both Bonaparte and Carnot understood this. As much is reflected in Carnot's initial instructions to General Bonaparte, which clearly articulated the need to remove the Piedmontese threat to French communications with the Republic.[30] The Austrians were another matter. Their generals were veterans, many of them having fought against Frederick the Great in the Seven Years' War as well as in the current conflict. Their greatest weakness was their caution and constant underestimation of the French and Bonaparte. But caution was also their great strength for it prevented Bonaparte on many occasions from trapping them and destroying their entire army. Only near the end of the campaign did Bonaparte finally destroy an army, and a small one at that, shut up in the fortress of Mantua, but it took him over eight months to do it. The Austrians were thoroughgoing professionals and their troops were veterans who fought well and sometimes out-marched the reputedly swifter French. A weakness at the operational level for the Austrians was the centralized control of their armies to a much greater degree by something known as the *Hofskriegsrat*, also known as the Aulic Council. This term stands for "court war council" and represented a sort of political-military general staff. Austrian generals had to get their plans of operation approved by this council and keep it updated any time major deviations occurred, as opposed to the French generals who often ignored Paris once the campaign had begun.[31]

Additionally, Austrian artillery, while not as standardized and light as the French, was only a little less well served—although the Austrians were not as adept as the French in moving their guns up close to support the infantry at the point of attack as the French were. Austrian regular infantry was composed primarily of the many storied imperial regiments known to history as the *Kaiserliche* (German for imperial). The grenadiers from these regiments were, like the French, often pooled together into grenadier battalions and columns and used as shock infantry. Many a time the French would receive a severe check from these fearsome elite warriors.[32]

Austrian cavalry was uniformly superior to the French cavalry of this period and when properly utilized excelled at reconnaissance, screening, and covering the infantry in retreat. Finally, the Austrians used various long-service light infantry formations such as the Croats and the frontier *Grenzer* regiments that enabled them to counter in many ways the swarms of French skirmishers and chasseurs that were swirling around a typical French formation, be it a battalion or an army. Their tactics conformed to the French and when the need arose, they could fight in line and column, too, often much more effectively. For these reasons, it is easy to see why it was the Imperial Austrian forces that alone remained in the field against the dauntingly large and energetic French revolutionary armies by 1796 and not the Prussians, Spanish, and all the rest.[33]

As mentioned earlier, Bonaparte's guidance was to achieve a short-term political end-state—the departure of the kingdom of Savoy from the war as soon as possible. Carnot and Bonaparte both believed that the many peace treaties and withdrawals from the First Coalition had rattled the king of Savoy. From his very first campaign, then, the young general clearly focused on using military operations for political ends—a defining feature of the operational art. He was then to turn and drive the Austrians eastward up the Po River valley and generally be as successful as possible in tying up Austrian forces in his secondary theater. This might prevent the Austrians from reinforcing their army in Germany, under Archduke Charles and

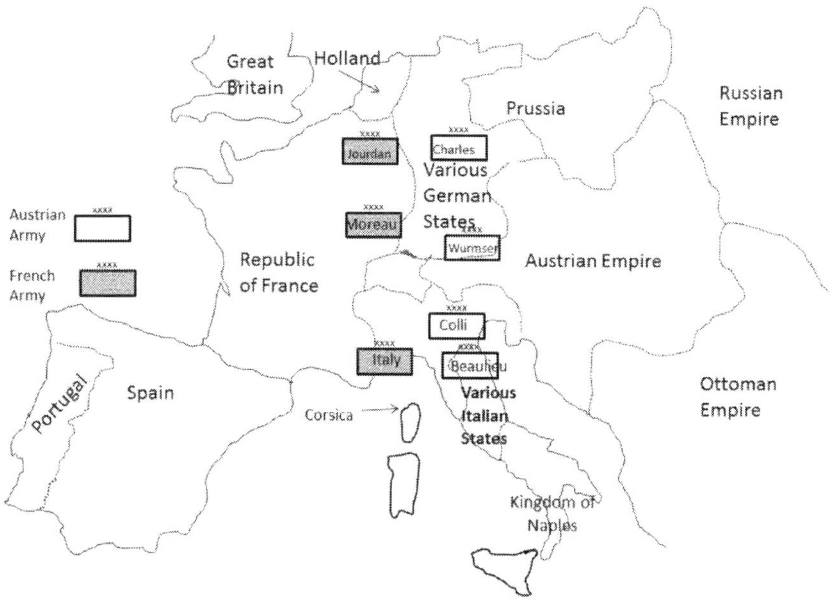

Map 1 – Strategic Situation 1796

against Generals Moreau and Jourdan, with over 140,000 troops into Germany from the lower and middle areas of the Rhine valley, respectively (see map 1). Bonaparte immediately went on the offensive with his starving and unpaid army—gambling that he could win and win quickly and then debauch into the rich Po valley where he and the representatives of the Directory could extort monies (especially hard specie) and find food. As it turned out, Bonaparte's army was a typical French army in this respect, down but not out and still full of fight despite its destitute character. War could be used to pay for war and underwrite a continuation of the offensive.

PHASE I: THE OPERATIONAL OFFENSIVE

The key to Bonaparte's operational design against the Austrian-Piedmontese coalition was to capture the central position in a key mountain pass that led over the Maritime Alps and separated the armies of Generals Colli and Beaulieu. Between them these two generals had over 55,000 troops as opposed to Bonaparte's threadbare field force of around 37,500. The Austrians helped Bonaparte to some degree by attacking the far right portion of his positions near Voltri in early April. After a perfunctory repulse by the local French commander, Bonaparte judged his right safe enough to launch his Italian "blitzkrieg" into the juncture of the two Allied armies. In 10 days in April, his forces won three major battles at Montenotte (April 12), Dego (April 14–15), and Mondovi (April 20) (see map 2). At the first location he surprised and mauled an Austrian division, and then defeated an Austrian counterattack at Dego. While the Austrians recovered, he turned on the Piedmontese forces and defeated them in detail, driving rapidly on the capital of Turin. On April 28, 1796, the King of Savoy (who ruled Piedmont and Sardinia) signed an initial peace with General Bonaparte, which was duly forwarded to Paris for ratification.[34]

This portion of the campaign represents the first phase of Bonaparte's operational offensive. It consisted of many minor engagements as well as the three main ones highlighted. These battles were often divisional affairs and in some cases consisted of the use of two divisions as Bonaparte moved his operational units fluidly and flexibly, always trying to retain the initiative and if possible overwhelm the enemy before he could consolidate with reinforcements or other friendly forces. Bonaparte's command and control relied heavily on the initiative of his divisional commanders, most of the tactical actions not being controlled by him, but by his veteran subordinates. His forte in this early campaign was in the way he moved his units around—one day the main effort being on his left, and the next that same unit being used as a

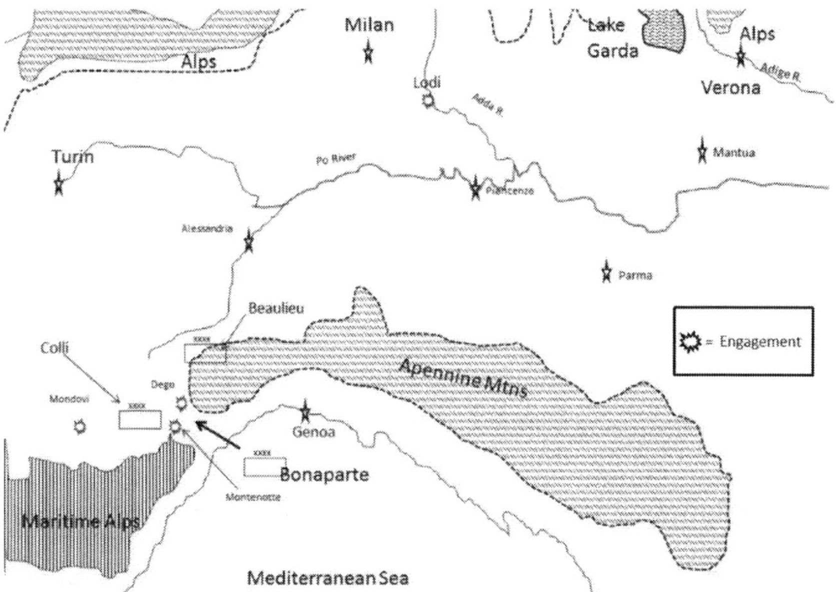

Map 2
Opening of Italian Campaign 1796

reserve or flanking force for another division. Bonaparte controlled at the operational level—but in order to sense how the battle went, he moved around the difficult terrain of this part of Italy, constantly on the move, writing orders, assessing the situation, and then issuing new orders as his computer-like brain integrated new information. Later writers, such as the air force officer John Boyd, would write extensively about Bonaparte's method as getting inside his adversaries' "decision cycle," using something Boyd characterized as the "OODA loop." OODA stands for observe, orient, decide, act—and his raw notes still attest to his fascination with how Bonaparte excelled at this sort of rapid decision-making at the operational level.[35]

The fluid nature of these operations reflects an early form of "echelonment" that is essential to the operational art. Instead of having fresh operational echelons available as unused reserves, Bonaparte often used the main effort of one day as the next day's reserve, echelon, or flank guard. For example, Masséna's division fought at Montenotte, but was then used in support of Sérurier's division in another part of the theater within a day's march of the recent engagement. Where this style ran into problems, as happened to Masséna at Dego, was that the poor logistics arrangements often resulted in unplanned "operational pauses" as the victorious, but famished and tired, French troops dispersed after a battle to rest or feed themselves. In Masséna's case, his division became a flank security force for the others, but was surprised as the Austrians returned to recapture Dego from the spread out, foraging French troops. Normally, cavalry serves the function here to screen, but as noted, the French cavalry was weak and often poorly led during this period, and as often as not was also out foraging (or plundering). Thus the Austrians retook Dego, which required another battle by a "new" French division to retake it again! However, the operational durability of the French divisions came through again and again in this portion of the campaign as the French divisions rapidly reconstituted and force-marched their way to some other sector of the expanding front.

On the negative side, the bulk of the Austrian Army under Beaulieu escaped and established a new line of defense along the Po River. Bonaparte had achieved his first military and political objectives, but to help out with the larger operational and strategic design for 1796 he had to continue to do what is known in military terms as "fix" Beaulieu's army in theater. This meant his job was to stay closely engaged and in proximity to the Austrians to prevent them from sending reserves to the northern theater. Put another way, he had to provide a credible threat to keep them honest. The second portion of this operational phase of the campaign did this in spades. Instead of merely observing and threatening Beaulieu, he boldly struck out both north and south of the Po, constantly outflanking the normally defensible north–south tributaries of that river as he made his way east with rapid outflanking movements. He did this based on his judgment that the Austrian general was still shaken and could be hustled even further back, perhaps allowing the French to accomplish the even bigger goal of acquiring the Austrian territories in Lombardy and moving the front well east and within striking distance of Austria itself.

Bonaparte gave Beaulieu no time to consolidate his defense along the Po. He opened his offensive in the first week of May, distracting the Austrians with two divisions at his bridgehead at Valenza (which belonged to the now neutral Piedmontese) while he sent a specially picked vanguard composed of four battalions of grenadiers culled from his regiments combined with his best cavalry, an ad hoc division totaling over 6,000 men under General Claude Dallemagne. This force marched before dawn on May 7 deep along Beaulieu's right flank along the Po. They arrived at Piacenza (see map 3) unexpectedly and crossed the Po by ferry unopposed. Bonaparte had echeloned the divisions of La Harpe and Augereau in support of Dallemagne, and La Harpe arrived just in the nick of time as an Austrian division arrived to attack the bridgehead. This move unhinged Beaulieu's defense and he immediately ordered his army to concentrate opposite the French bridgehead at Piacenza.[36]

Now an operational pursuit began, with two divisions north of the river pursuing the retreating Austrians and the equivalent of two other divisions on their flank and rear expanding their bridgehead. A confused night meeting engagement occurred at Cadogno that saw La Harpe killed by friendly fire, but the encounter so unsettled Beaulieu that he made the critical decision to abandon Milan and retreat across the Adda River line through Lodi toward the safety of the fortress of Mantua. Bonaparte's hard marching divisions now united with each other in a race to catch the Austrians before they crossed the river, but again the cautious Austrians had withdrawn their neck from the noose that Bonaparte had so carefully prepared. Driving hard, Bonaparte's vanguard, followed again in echelon by Masséna and Augereau, caught up to the Austrians at the bridge over the Adda River at Lodi where a strong rearguard of 10,000 troops was well placed to contest the crossing. Bonaparte made the snap decision to try and add to his operational victory by scooping up the Austrian rearguard. Initially his forces failed, but he led them in person into the fray and they crossed both the bridge and a ford downstream and took the Austrians in the front and flank. At some cost, he bloodied the Austrians, who lost three times as many men as the French as well as16 guns, but the Austrians withdrew the bulk of their force in good order, covered as always by their excellent cavalry. Five days later, Bonaparte arrived in Milan to an enthusiastic crowd that greeted him as a liberator from the Austrian yoke.[37]

Bonaparte's operations in this busy first week of May highlight again his rapid style of maneuver and decision, bluffing as often as bludgeoning his enemy out of a very strong position. The march of the vanguard in concert with operational deception in crossing a contested river line has few equals in military history—and Bonaparte would not equal this feat again until he crossed the Danube River near Vienna, under grave circumstances in 1809. The deep march of the vanguard had all the hallmarks of an embryonic deep operation as envisaged by the Soviets like Triandafillov.

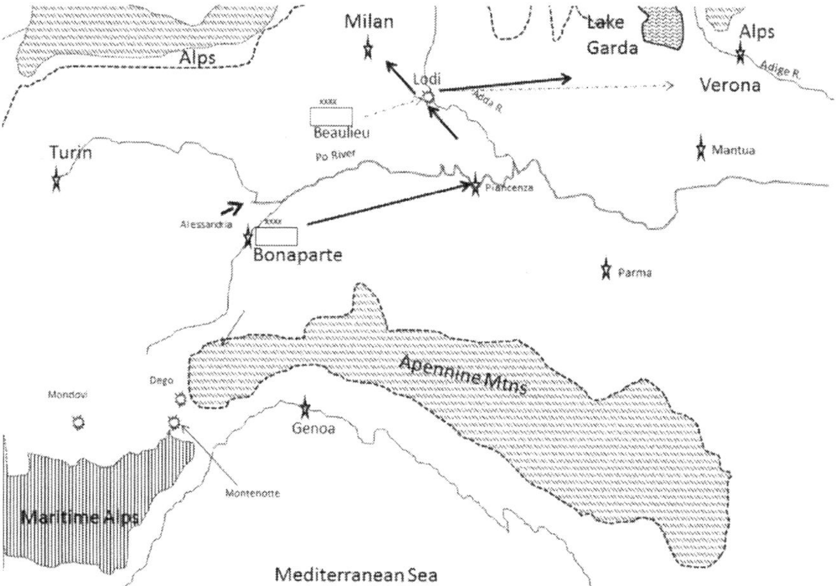

Map 3
From Lodi to Verona 1796

Even the depth, over 30 kilometers, matches with what the Soviets anticipated for deep operations by the first echelon operational force. Bonaparte's use of echeloned divisions, something that would become standard later for his corps commanders, is also worth noting.

In the aftermath of the victory the Directory tried to bring General Kellermann, the hero of Valmy, into the theater to command a second Italian army while they sent Bonaparte down the Peninsula to terrorize the Pope and the Kingdom of Naples, which was still at war with France, albeit half-heartedly. Bonaparte, now the victor of Lodi and the liberator of Milan, objected and the government gave way—sending most of Kellerman's troops to Bonaparte as reinforcements. Kellerman even sent his son, a cavalryman, to serve under Bonaparte in subsequent operations. The younger Kellermann became one of Bonaparte's most competent and fearless cavalry commanders. In this instance, Bonaparte was probably correct in demanding he retain the sole command in Italy—the division of command in the north had only resulted in defeat and disappointment. As Bonaparte commented, "One bad general is better than two good ones."[38]

In the aftermath of Lodi, the French consolidated their position, put down some local rebellions, and then pushed on toward Mantua in Venetian territory. Beaulieu attempted to defend again along the Mincio River, the last obstacle prior to Mantua. Again, Bonaparte feinted at one spot and crossed successfully at another. Beaulieu retreated with his beaten army to north toward the Tyrol after leaving a sizable force in fortress Mantua as a garrison. After a failed coup de main against the great fortress on May 31, the French settled in for a siege. Bonaparte's forces also secured their southern flank with the Papal States and Tuscany, essentially blackmailing them to keep the French from invading and looting these territories. For the time being French offensive operations ceased and they were much consumed with pacifying their new Italian conquests, which soon tired of their rapacious French "liberators."[39]

PHASE II: THE OPERATIONAL DEFENSIVE ON INTERIOR LINES

Bonaparte assumed the operational defensive, covering the siege of Mantua. However, because of the anemic showing by the French armies in Germany, the Austrians were able to reinforce and reconstitute Beaulieu's army under a much more combative and competent commander, Dagobert Sigismund von Würmser. With 50,000 troops, Würmser decided to advance against the French forces besieging Mantua in three columns. His hope was that he would unite them in the vicinity of Mantua and lift the siege. With the garrison of 12,000 troops added to his army, he then planned to retake Lombardy. One column each would advance along the east and west shores of Lake Garda with a third column advancing down the Brenta River valley (see map 4). Würmser personally accompanied the central column advancing down the eastern shore of Lake Garda. Bonaparte had to honor all three threats, and if he fought one, the other two would unite on his flank or rear and overwhelm him—at least this was the Austrians' hope.[40]

When Bonaparte finally gained solid intelligence of Würmser's advance in late July, he abandoned the siege of Mantua and moved north to face the threat. Initially, the Austrian columns advanced steadily, forcing the French back at every point. However, the coordination of the three columns offered Bonaparte an opportunity to again strike out from a central position at the southern end of Lake Garda. Würmser's westernmost column under Quasdanovitch was rudely handled by Masséna's division as it attempted to unite with Würmser's lead elements at Lonato on August 3. At the same time, those lead elements from the central column were turned back within six miles of Lonato by Augereau in the vicinity of Castiglione.[41]

On August 5, Bonaparte brought two divisions (including Masséna's) to assist Augereau facing Würmser at Castiglione. Around 30,000 French troops faced 25,000 Austrians; Bonaparte had again managed to mass more men and bring on an engagement under favorable circumstances. He had to settle for a tactical victory, rather than his larger design of annihilating the entire Austrian force. Würmser, a skilled cavalry

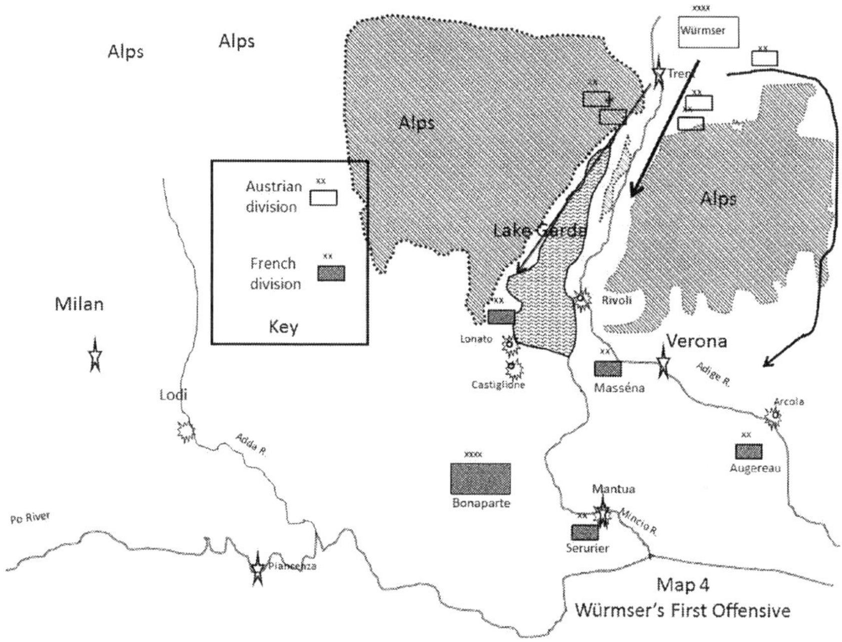

commander, managed to get the bulk of his force away, although he lost 20 cannon. He gathered up the division from the third column (General Meszaros) near Mantua after slipping reinforcements and supplies into that fortress and then withdrew rapidly up the valley of the Adige to Roveredo. In addition to the losses sustained by Quasdanovitch, he had suffered only about 3,000 total losses (including prisoners taken during his retreat). He still had the bulk of his army intact, if somewhat rattled by the Corsican prodigy's rapid maneuvers and propensity for engagement. But the old Austrian had one column that was relatively fresh and unengaged and was willing to try a second offensive just as soon as he could restore order to his forces.[42]

Meanwhile, in Paris, the Directory chased events in the field, first directing Moreau to advance to relieve the pressure on Bonaparte, and then, hearing of Würmser's repulse, directing Bonaparte to attack toward the Tyrol to relieve pressure on the northernmost French army under Jourdan being pressed by Archduke Charles. Meanwhile, after receiving more reinforcements, Würmser began his second offensive, now concentrated into two forces—one protecting the Tyrol in the vicinity of the northern end of Lake Garda under Davidovitch (20,000) and another under himself of about the same strength. His plan was simple, while Davidovitch kept the French engaged and protected his rear, he would advance rapidly down the Brenta valley, then relieve Mantua and, hopefully, crush the Army of Italy in the Adige valley (see map 5). This maneuver scheme required very rapid movements by the Austrians, a solid performance by Davidovitch, and did not account for anything Bonaparte might do. It was to be deficient on all three counts.[43]

Davidovitch had been charged by the *Hofkriegsrat* to protect the Tyrol, and Bonaparte's initiative paid wonderful dividends as he ignored Würmser's wide flanking movement down the Brenta valley and launched a spoiling attack with the divisions of Vaubois and Masséna against Davidovitch. The cautious Austrian was hustled north by a series of engagements in early September and Bonaparte again gained control of a central position between the major forces of his enemy. Leaving Vaubois to screen Davidovitch, he took the bulk of the French army and crossed the mountains into the upper valley of the Brenta and rapidly approached Würmser's rear. Bonaparte overran Würmser's rearmost forces on either side of the Brenta at Bassano on September 8. Meanwhile, the lead Austrian division was unable to cross the Mincio River at the fortress of Verona and Würmser frantically recalled it to the east to join him. Bonaparte cut off Würmser's line of retreat, causing Quasdanovitch's spent division to retreat eastward toward Trieste. The French were within a hare's breath of trapping and annihilating Würmser's dispirited army against the Mincio River, but

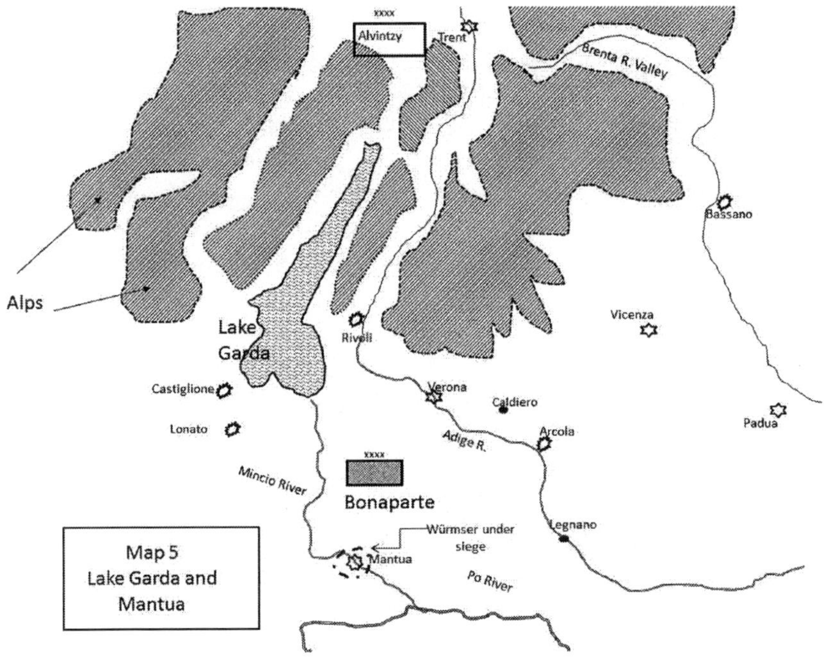

the old fox managed to slip across at Legnano and retired into the fortress at Mantua. Bonaparte's exhausted army arrived a few days later to reestablish the siege.[44]

Napoleon Bonaparte had not destroyed the Austrian army, but he had come very close and trapped Würmser and 24,000 troops in Mantua. With this many troops in so confined a space, disease soon broke out and the number of effective troops declined due to death, privation, and hunger. Würmser would surely capitulate unless the Austrians could raise another army to liberate him. The Austrians, with both Moreau and Jourdan defeated along the Rhine, did just that. Moving troops from the German theater and using what remained of Würmser's army (Davidovitch and Quasdanovitch), the *Hofkriegsrat* formed another powerful force to drive south and relieve Mantua and end the string of French victories by the upstart Corsican general. These troops were under the command of Baron Joseph D'Alvintzy. Alvintzy's plan bore a strong resemblance to Würmser's earlier plan—the main effort under Alvintzy would advance down the Brenta River to the fortress of Verona while the forces in the Tyrol—still under Davidovitch—moved down the valley of the Adige River to join the main force in a pincer movement at Verona. The chief difference between the two situations was that this time Bonaparte had a serious military threat under a very competent general in his rear at Mantua. His challenge would be to mask Würmser with a large enough force while using a much smaller force than previously to fight off Alvintzy and Davidovitch—it was to result in the greatest challenge of the campaign. He would be outnumbered by over 20,000 men when one includes the garrison of Mantua in the Austrian order of battle.[45]

At the insistence of the *Hofkriegsrat*, Alvintzy's offensive began in early November. Bonaparte's plan to meet it involved opposing both advances with single divisions. He also had to maintain a strong operational reserve around Verona under himself. He waited for an opportunity to strike the Austrians separately before they could unite in the vicinity of Verona. Davidovitch's column performed its portion of the task effectively, manhandling the French division facing it and pushing down the Adige valley toward Verona. Fortunately for Bonaparte, Alvintzy advanced lethargically against Masséna's division. Bonaparte rushed to shore up his left flank, bringing Masséna with him (but not his troops) and restored his shaken troops. At the same time, Davidovitch paused, giving the French some extra time to rest and reconstitute after their earlier defeat. Alvintzy, meanwhile, had resumed his advance but did not have the bulk of his army with him. Bonaparte, with a wary eye on his rear, gathered his "operational reserve" and proceeded to Caldiero where he attacked Alvintzy's 18,000 troops on November 12. Bonaparte failed to throw Alvintzy back, taking more casualties than he inflicted. He left a small garrison in Verona and retired to the western side of the Adige River line. His plan was to recross the river to the south and seize the Austrian supply trains, forcing them to fall back through lack of logistics. The result was a hard-fought, three-day battle in the marshes and streams around the

Adige in the vicinity of Arcola (November 15–17; see map 5). Bonaparte assumed the tactical offensive, attacking again and again with his smaller force (initially 19,000 strong). Both armies suffered greatly, but Alvintzy's nerve gave way on day three and he retreated up his line of communications to the Brenta valley.[46]

Through hard fighting and some luck, Bonaparte had survived the key crisis of the campaign, although Alvintzy's army was still "in being." As to the other Austrian forces and Bonaparte's luck—Davidovitch had remained lethargic during much of the battle and Würmser did not sortie. Bonaparte, showing some sophistication in joint warfare, had built gunboats to help patrol the swamps surrounding Mantua and cover the exits from which a sortie might erupt. Bonaparte now punished Davidovitch for his lethargy, gathering up the garrison of Verona and most of his remaining army he rushed north to the vicinity of Rivoli. Davidovitch realized his peril and scuttled back up the valley toward the Tyrol.

The Austrians were still not finished. Despite this latest setback, Vienna demanded another push to relieve Mantua, and Alvintzy gathered over 28,000 troops to descend the Adige valley through Rivoli to do just that. He left considerable detachments to perhaps break through to Würmser while he fought Bonaparte in the north. This plan failed miserably when Bonaparte, on the tactical defense at first, repulsed Alvintzy at Rivoli on January 14, 1797, and then achieved perhaps his greatest tactical victory to date the next day, taking the offensive and eliminating Alvintzy's army as an operational threat, inflicting 15,000 casualties upon it. Next to Arcola, it was the most decisive French battle of the campaign, although he had, as usual, been very ably supported by his subordinates, Masséna in particular. Again Bonaparte's central position and active defense helped him gather even more laurels as his men repulsed a sortie by the weakened garrison of Mantua and then cut off one of Alvintzy's relief columns and forced it into the already overcrowded fortress. These final operations sealed Mantua's doom and Würmser finally surrendered on February 2 to spare his men any further suffering and death.[47]

PHASE III: THE OPERATIONAL OFFENSIVE—AGAIN

With Mantua secured, Napoleon Bonaparte now went over to the offensive. He had made his theater of the war the main theater, and the Austrians now moved troops they felt they could spare to the Italian front to meet the crisis. They also rushed their best general, Archduke Charles, south to size the situation up and take command. Many generals might have rested and consolidated their gains, but not Bonaparte. He gave Charles very little time to reconstitute the demoralized Austrian troops and integrate the reinforcements that seemed to trickle in from Germany. In the meantime, the Directory sent substantial reinforcements to Bonaparte (two divisions), crafting a new strategy to make his the main effort while at the same time going on the offensive in the now-secondary theater of Germany with the armies there, pressuring the Austrians at all points.[48]

Meanwhile, Bonaparte, under orders from the Directory, led a force of his freshest troops, if they might be called that, against the Pope and forced the pontiff to sign a peace treaty in mid-February while at the same time extorting some 30 million francs and gaining key territories in the northern Papal States to add to the expanding French holdings in Italy. Even with these distractions and deletions, Bonaparte was able to launch a renewed offensive in early March against Charles's dispersed forces on a broad front, sending the young and talented General Barthelemy Joubert and a prototype corps of two divisions with attached cavalry against the Tyrol while he advanced against Charles's outposts along the Piave River (see map 6). Austrian imperial possessions were thus directly threatened in three critical areas—the Tyrol, Dalmatia, and Carinthia—the tables had turned, it was now Bonaparte forcing his enemy to prioritize his defense. At the same time, the French armies in Germany, now under Hoche as well as Moreau (Jourdan had resigned), also applied pressure that pinned significant Austrian forces down.[49]

Charles planned to try to defend along the Tagliamento River line (see map 6). Bonaparte further divided his force, using the mountain-fighting Masséna and his division on his left flank to threaten the Austrian line of communication to Carinthia

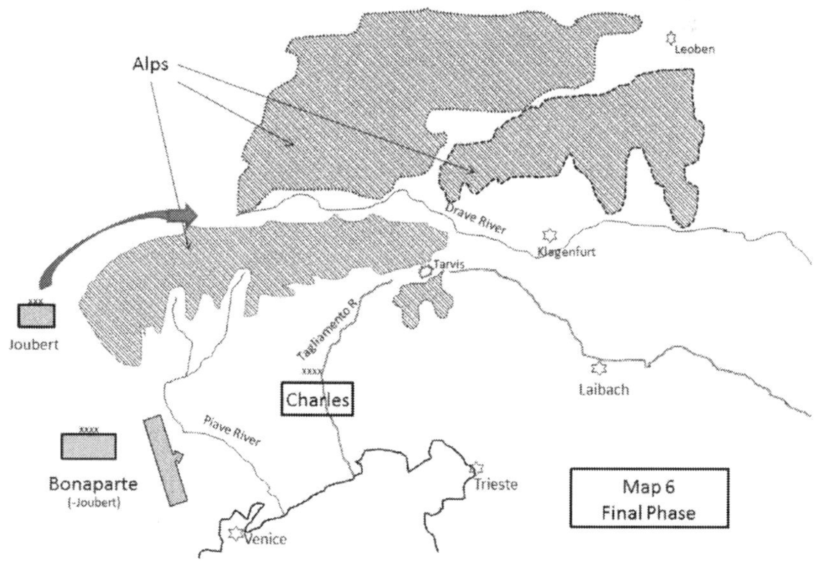

through the great pass at Tarvis. Many small rearguard engagements occurred along the route of the two primary columns, and Charles was driven steadily backward, never adequately establishing a defense that could not be outflanked. Bonaparte additionally peeled off another division to threaten Trieste and thence Laibach, which caused Charles to detach more troops from his main force in order to honor the threat. Unlike Bonaparte, he seemed unable to move quickly enough to smash these flanking forces. Joubert advanced in the Tyrol, winning a battle at the Avisio River on March 20, although he began to encounter problems with Tyrolese irregular forces along his communications. Two days later, Masséna's advanced guard arrived at Tarvis but was pushed out by a furious Austrian counterattack. The fierce battle continued as Bonaparte's divisions advanced on Tarvis from the south and he squeezed Charles's forces between his own and Masséna's columns. His forces defeated and captured Charles's rearguard of 2,000 troops and 25 cannon, adding to a toll by this point of over 15,000 casualties inflicted on Charles's army. Once again, the French commander never allowed the Austrian archduke to stabilize the front.[50]

United with Masséna, Bonaparte now pushed on toward Klagenfurt. Joubert, whose lengthy line of communications with Verona had been cut by the Austrians abandoned these and came east via the Drave valley to join the main army for the final drive toward Vienna. At the same time, Bonaparte also pulled in the division of General Jean Bernadotte up from Laibach, concentrating at Klagenfurt and offering the Austrians an armistice while continuing the offensive in early April toward Brunn and Graz. The Austrians were now eager for peace with the French so close to Vienna and their armies retreating in Germany. By April 7, Bonaparte was negotiating with the Austrian representatives after signing a truce. Bonaparte now proceeded to demonstrate his considerable diplomatic skills, negotiating a preliminary treaty on behalf of France and signing it on April 18 at Leoben. This peace was later formalized by the Treaty of Campo Formio that October, which formally ended the War of the First Coalition. France was given a free hand in Italy and Switzerland while Austria received Venice in compensation. The treaties also recognized the French acquisition of Belgium and German territory on the west bank of the Rhine.[51]

* * *

The 1796–1797 campaign emphasized many elements of operational art as defined in chapter 1. The durability of the units resulted in a sustained and lengthy campaign for over a year. Also, Bonaparte had resorted to a protracted war of attrition against the repeated Austrian attempts to reclaim their lost position in northern Italy. The division was the main unit of operational maneuver in the campaign, and not just by the French, although the French divisions tended to be more mobile than the Austrians and reconstituted more quickly, in most cases, than their counterparts. This was also due to the French losing, generally, much fewer troops in the many engagements as prisoners and casualties. Another feature of operational art is the constant combat and engagement that typifies the more modern type of combat, and this campaign, with some exceptions, was really decided by successive and constant engagement. As noted, on several occasions, Bonaparte's skillful use of maneuver resulted in his being able to win engagements by echelonment, as suggested by later operational art theory.

The fluidity of the campaign, even today, can make one's head swim—the lengthy marches and maneuvers by both sides were really quite remarkable given how difficult the terrain was and the numbers of significant river and alpine obstacles that impeded maneuver in the old style. Although there was no instantaneous communications, Napoleon seemed to have just the right touch—a type of mission command as mentioned earlier—with his subordinates that served him well when they were on their own without specific guidance from above, making timely decisions based on local circumstances. The Austrians fared less well in this respect, fighting much more in the old style—as in the situations where Würmser and Alvintzy could have profited mightily from more energy by their subordinates in independent command.

Finally, both Austria and France, during various points in the campaign were very flexible in moving operational reserves into the theater across the Alps from Germany when conditions permitted. However, once Bonaparte was given the substantial reinforcements he needed after the capitulation of Mantua, he was able to resume the offense and recapture the operational initiative and this took him all the way to Leoben. At Leoben he demonstrated a rare skill, the ability to take the operational fruits of the campaign and translate them into a favorable political result, something not often given to the province of generals. Napoleon Bonaparte's actions suggest that generals must also be well versed in politics and diplomacy to properly practice operational art, otherwise all the elegance and skill in producing a masterpiece campaign could be undone, either with a continuation of the war or a less than satisfactory peace. Some historians have argued that this latter result is precisely what the French got in 1797 at Leoben and then Campo Formio—but certainly the end of the War of the First Coalition and France's reduction of enemies to just Great Britain placed her in a very advantageous position.[52]

THREE

COALITION OPERATIONAL ART IN 1799

SUVOROV AND MASSÉNA IN ITALY AND SWITZERLAND

CHRONOLOGY

October 1797	France and Austria signed Treaty of Campo Formio
Jul 1798	Napoleon led an army to capture Egypt
Aug 1–2, 1798	Admiral Nelson destroyed French Fleet at the battle of the Nile
1799–1801	The War of the Second Coalition
March 1, 1799	French attacked Austrians in Germany and Switzerland
March 7	Masséna won battle at Chur
March 25	Archduke Charles defeated French at Stockach
March 26	Masséna captured Martinsbruck in Austrian Tyrol
April 5	Austrians won Battle of Magnano near Verona
April 14	Marshal A. V. Suvorov took command of the Allied army in Italy
April 26–28	Suvorov's army forced Adda River line
April 29	Allies "liberated" Milan
June 3–4	Masséna repulsed Archduke Charles at the First Battle of Zurich
June 17–19	Battle on the Trebbia River, Suvorov defeated Macdonald
July 27	French garrison of Mantua surrendered to Allies
August 15	Allied army under Suvorov smashed French at Novi

August 17	Archduke Charles began marching north to Germany
August 25	Suvorov and Russian army ordered to Switzerland
September 25–26	Masséna won the Second Battle of Zurich
October 5	Suvorov retreated over the Alps
November	Bonaparte returned to Paris a hero after abandoning his army in Egypt and established himself as First Consul
October 22	Tsar Paul withdrew Russia from the Second Coalition

Operational art executed on behalf of a single political entity, whatever its many goals, is simpler than in an alliance or coalition construct. In chapter 2 concerning 1796–1797, the operational artist Napoleon Bonaparte had only one constituency to satisfy with his "work"—the government of France. In coalition warfare, the situation is much more demanding and complex for the operational artist—he or she must satisfy more than one political master, although certainly the priority goes to the state from which the general originates or to whom he has sworn service (not always the same in that era of cosmopolitan 18th-century generals). For example, Dwight D. Eisenhower served U.S. interests in World War II while at the same time trying to also satisfy the political interests and agendas of his sometimes troublesome allies, the French and the British. The point to be made here is that when campaigns are designed to accomplish military objectives that lead to political results, this is often more difficult in a coalition construct than without it. Napoleon's foreign minister, Charles-Maurice de Talleyrand, derided coalition warfare during the period (referencing Prussia and Austria): "We [French] have given up fearing coalitions: there is a principle of hatred, jealousy and distrust between the Cabinets of Berlin and Vienna which will guide them above all else."[1]

Into this difficult mix arrived the towering personality of Marshal A. V. Suvorov in command of an Austro-Russian coalition army. Suvorov had fought the Prussians in the Seven Years' War and had been one of Catherine the Great's most successful generals against the Turks and the Poles.[2] At this time his reputation as a great general was well established, being on par with Napoleon Bonaparte, Saxe-Coburg (whom Suvorov had asked be appointed to command in Germany), and Archduke Charles. Suvorov was charged with nothing less than reconquering Bonaparte's Italian gains for Austria while that general remained cut off with the French expedition to Egypt.[3] Suvorov is of particular interest because later Soviet historiography and military writing identified him as a sort of prototype for the ideal Russian operational artist—especially his very advanced ideas on the importance of training troops.[4]

Secondarily, the operations of *Général de Division* Andre Masséna in Switzerland will also be examined. Masséna's operations are critical to understanding the flow of operations in Italy, but ultimately Massena's successful campaign in Switzerland, along with a disastrous Anglo-Russian campaign in Holland, were major causes of

Russia leaving the coalition and the war. Finally, Masséna's operations provide further evidence of an operational artist hard at work against a superior foe, mixing offense and defense in the right amounts to overcome difficult challenges and in no small measure prevent the invasion of France by victorious Austrian and Russian armies.

THE CAUSES OF THE WAR OF THE SECOND COALITION

As mentioned in chapter 2, the events that led to the outbreak of war had their roots in the incomplete peace agreed to by the French and Austrians at Campo Formio in 1797. However, French actions and provocations played an equally important role. Also, to understand the operational-level interactions, it is necessary to grasp the political context for the fighting, especially in the Italian theater of operations. The Second Coalition has been much criticized—yet, as we will see, it came very close to success. Of all the anti-French coalitions formed before 1813, it superficially resembled the successful coalition that emerged 14 years later in 1813. The coalition armies in both contests outnumbered the thinly stretched French forces at the beginning of the respective campaigns. As in 1813, a significant reason for the lack of veteran cadres for a continental contest would be Napoleon Bonaparte, cut off in Egypt with the cream of the French army in 1799 and by his own disastrous retreat from Russia in 1812 (see chapter 7). General war weariness and disillusionment with the French "liberators" in the European territories they now "defended" was also a common trait for both 1799 and 1813. Similarly, rebellions in Naples and Switzerland foreshadowed the mix of conventional and guerilla warfare later in Spain, Russia in 1812, and Germany in 1809 and 1813.

Austria signed the Treaty of Campo Formio in 1797 but Great Britain continued to fight because of French domination of the Low Countries and the resultant closure of these ports to British trade. Also, the specter of a unified Franco-Dutch fleet posed an even greater nightmare for the politicians in London than the economic predicament (see chapter 4). Great Britain redoubled her efforts to find continental allies, but for all intents and purposes she remained alone against France. The Franco-Austrian peace followed a common pattern for treaties during the entire Napoleonic era: it included both open and "secret" provisions. The secret provisions allowed the French and Austrians to get around other obligations that they had incurred in previous treaties, making for *realpolitik* diplomacy in this period as cynical as any in the modern era. The secret clauses in Campo Formio included recognition of the Rhine as a "natural boundary" for France, yet the details of this boundary remained unresolved. These continuing issues along the Rhine and Austria's foothold in Italy presaged a likely conflict in the future. French revolutionary leader Abbé Sieyès summed it up best, "This treaty is not peace: it is a call to new war."[5]

Great Britain regarded these treaties as outright betrayal. To make matters worse, Austria defaulted on her loan convention with Great Britain. This one issue was to poison the atmosphere between the two countries that needed to cooperate most in their war against France. Austria was under no illusions about a future conflict with France. As a result, the two most inveterate foes of revolutionary France barely communicated with each other while their respective leaders negotiated for the creation of a new coalition. The dispute came to a head in May 1797 when the Austrian Foreign Minister Baron Thugut refused to ratify a convention for the repayment of Austria's war loans from Great Britain. Later, when Austria approached Britain in April 1798 with an alliance proposal, it foundered on the rocks of the unratified loan convention. Britain would coordinate, rather than cooperate, with Austria in the war against the French.[6]

Against this backdrop the French continued their territorial aggrandizement under the guise of peace. French aggression not only provoked an outraged Europe into forming a second coalition but also had the additional effect of further weakening France by overextending her military. The first act of military overextension was Bonaparte's expedition against Egypt. His purpose was to strike at the British lines of communication with India. One very significant action incidental to this expedition was Bonaparte's seizure of Malta from the Knights of St. John. Unfortunately for the French, Russian Tsar Paul I was the self-proclaimed protector of this order and aspired to the title of grand master. The Tsar had already been in serious negotiation to field an army against the French. The French conquest of Malta convinced the Tsar to join the ranks of France's enemies. The other key aspect of the French aggression in the east was the addition of Turkey to the ranks of France's foes. On September 5, 1798, the Turks allowed a Russian fleet "for one time only" to pass through the Bosporus.[7]

Bonaparte was not alone among the French in his ability to add enemies to the field against the Republic. Before he departed for Egypt he was already cognizant of the Directory's plans for new aggression against the Papal States, the United Provinces (Holland), and the Swiss federation. One historian articulates this policy as the "Napoleonic practice of using peace as an extension of war."[8] Two events now occurred in Italy as a prelude to the coming war. Each, in its own way was to affect the Second Coalition's conduct of operations. The first occurred in the Bourbon Kingdom of Naples. Egged on by the British admiral Horatio Nelson, the hero of the Battle of the Nile, the Bourbons sent their army to "liberate" Rome in December 1798. The French dispersed the Neapolitan forces and then proceeded south and displaced the Bourbon King, conquering his country and establishing the satellite Parthenopian Republic. In a second action, the French convinced the King of Piedmont, Austria's erstwhile ally, to abdicate to Sardinia. Upon his departure, the Piedmontese army, with his consent, was incorporated into the French forces in Italy.

The issue of Piedmont was to cause serious problems for the Second Coalition during Suvorov's campaigns in 1799.[9]

In summary, the French had to maintain armies of occupation in Holland, Italy, and Switzerland, to say nothing of Egypt. The situation in Naples was particularly onerous. The French occupation had resulted in a wave of patriotism and resistance by the *lazzaroni* (beggars) in the cities and the *sandfesiti* (peasants) in the countryside.[10] General (later Marshal) Etienne-Jacques Macdonald, the commander in this sector, commented that "no sooner was insurrection crushed at one point than it broke out in another."[11] British sea power, too, demanded garrison troops along a lengthy, vulnerable coastline that now included Belgium and Italy. Additionally, the threat of a renewed Austro-Russian onslaught in Italy or along the Rhine (including Switzerland) caused the French to maintain substantial armies in response. The Directory prepared to face the armies of the Second Coalition under these considerable constraints. Instead of contracting or withdrawing its forces, the French government adhered to the "principle of keeping everything, and not yielding a foot of ground," as Macdonald bitterly observed.[12]

The end of the eventful summer of 1799 saw the Second Coalition poised to achieve the strategic success denied the previous anti-French coalition.[13] Indeed, the initial campaigns resulted in both tactical and (by modern standards) operational victory. British sea power had cut off and stalemated General Bonaparte in Egypt while the Archduke Charles defeated or neutralized the French, under Generals Jourdan and Masséna, in southern Germany and Switzerland, respectively. Finally, the Russian field marshal Alexander Suvorov had conquered Italy at the head of a combined Austro-Russian army. Suvorov's campaign was in many ways as brilliant, and certainly shorter, when compared to Napoleon's more celebrated 1796–1797 Italian campaign. Suvorov and his Austro-Russian army effectively destroyed one French army (Scherer and Moreau), severely defeated another (Macdonald), and crippled a third (Joubert). Suvorov "liberated" almost the whole of Italy, placing France territorially farther back than the starting point of Napoleon's 1796 campaign.

Yet, before the year's end this promising situation had imploded. The Second Coalition collapsed as the Russians withdrew in disgust and became hostile neutrals.[14] The final victory that had seemed a certainty in September was doubtful by December. What caused this dramatic and rapid turnabout? In order to understand what happened, and how the brilliant operations of Suvorov in Italy came to naught, we must go back again to the start of the campaign in early 1799.

One must first examine the raw material of Suvorov's armies—the Russian and Austrian soldiers. Understanding the Russian soldier of this period is the key to understanding the corresponding operations that Suvorov developed on his own initiative. Sir Robert Wilson, the British liaison officer who served extensively in Russia, observed that the Russian soldier "is fearless, disdains the protection of ground, is *not intimidated by casualties* [italics mine]."[15] Archduke Charles thought the "officers and men were poorly trained in tactics and deployments." Against European armies the Russians normally employed the 18th-century linear formations and against the wilder Turks and Poles they employed columns and squares.[16] As for skirmishing, Russian light infantry (*jaeger*) formations did exist, but skirmishing and independent actions by them were only in the regulations and still in their infancy in terms of actual practice.[17] This, then, was the Russian army in 1799: stolid, imperturbable infantry, numerous artillery indifferently served, and clouds of Cossacks. Combined arms action was virtually nonexistent, except, as we shall see, when Suvorov was in command.[18] The émigré Frenchman Count Langeron—who was one of the Tsar's many foreign officers—captures this essence best: "All their principles of war come down to their bayonets and their Cossacks . . . and all their enterprises have been crowned with success."[19]

Their leader in this campaign was truly a "soldier's general." Hard on his officers, Suvorov was paternal with his men, caring for their well-being, eating their fare, and sleeping on hay or the ground as they did. He shared their hardships and they idolized him in return. One historian goes so far as to say that only Suvorov, of all the Russian generals before and since, really understood the "potential for . . . innovation that the Russian peasant soldier afforded" his leaders.[20]

Another key to understanding Suvorov's success is to understand his revolutionary methods of training. In simplest terms, Suvorov trained for both the tactical and operational environments. His motto was "train hard, fight easy." He wrote: "Training is light, the lack of training is darkness . . . the trained soldier is worth three soldiers who are untrained. No! it's more than that, say six, still not enough—ten is more like it."[21] He took the basic tactics of the Russian army, modified them to his purposes, and then ruthlessly *practiced* them with the troops under his command. At the operational level he routinely had his troops make forced marches *in peacetime* of 26 miles or more to build up their endurance. His soldiers often marched as fast, or even faster, than the average French Revolutionary armies. Perhaps one key to this sort of performance was simply that the other Russian generals did not possess Suvorov's drive, energy, and impetuousness—character traits necessary for the true operational artist. His practice of combined arms maneuvers prior to engagements is justly famous. He had infantry charge infantry with bayonets leveled and cavalry charge full tilt against infantry who passed them in prearranged corridors only at the last possible second. He also had his cavalry in formation near his cannon when they fired to condition the horses not to panic from this source. Suvorov had his men conduct mock attacks on monasteries to simulate the stone fortresses of Europe.[22]

Suvorov also practiced what Jomini later called grand tactics, which is the art of arranging and deploying troops on a grand scale for battle prior to actual combat. However, he differed from Jomini by having his troops practice the key transition period between the approach march (operational maneuver) and the actual conduct of the engagement—always tailoring his tactics to suit whichever enemy he faced, whether a formal European army or the more irregular forces of the Poles and Turks (and French). He was particularly adamant that his troops be able to deploy directly from the march into formations based on the terrain. He did this because he was the master of a type of engagement most generals avoided; the meeting engagement or "accidental" battle between two armies on the move and not formally deployed for battle. For Suvorov, though, because of his method, these engagements were almost always deliberate. Bonaparte would later employ the same method with his *corpsd'armée*. The French had become the reigning masters of this sort of battle with divisions that deployed rapidly into skirmishers, lines, and columns, but they were to meet their match in Suvorov. He practiced a kind of Russian version of John Boyd's "OODA" cycle (see chapter 1), using *surprise, assessment*, and *speed*—what one historian calls his "triad"—to turn the operational and tactical into one dynamic, violent continuum. He trained his soldiers accordingly, making them understand that surprise was normal by introducing it into his training exercises.[23]

Suvorov explained this method during the 1799 campaign to the Austrian general Bellegarde: "speed and surprise substituted for numbers [while] hitting power and blows decided combat." He got his striking power from his constant focus on training. Like his counterpart at sea, Lord Nelson, he sought rather than eschewed battle in order to apply that power directly to his surprised enemies. On land he materialized much as Bonaparte had done from unexpected directions and then ruthlessly kept his enemy fixed as other echelons arrived to contribute to his combat power. What is most fascinating about this campaign, his last, is that he obtained a nearly similar result with his Austrian troops that he had not trained. This was a testimony to his methods of leadership but also to the inherent untapped potential of the veteran Austrian troops he employed as part of his army.[24]

After its numerous defeats during the War of the First Coalition (1792–1797), the Austrian army had attempted to implement new reforms. This effort was first placed in the hands of Archduke Charles, who counseled only minor changes because of the imminent outbreak of another war with France. Charles wrote, "before such changes have been completed for some time, the army will be in disarray and there will be disaster." Charles was removed and a more pliable military commission replaced him, headed by the same General Alvintzy defeated so handily by Bonaparte at Arcola and Rivoli.[25] In the end, the only "reforms" implemented were a new musket, new uniforms, and a superficial reorganization of the army. A good feel for the doctrine is captured in the following excerpt from the book *Generals-Reglement* of 1769:

> Regular, trained, and solid infantry, if it advances in closed ranks with rapid steps, courageously, supported by its artillery, cannot be held up by scattered skirmishers.

> It should therefore refuse to lose time either by skirmishing or by the fire of small groups. . . . It should close with the enemy as rapidly and orderly as possible, so as to drive him back and decide the action quickly. This is the method that *saves lives*; firing and skirmishing *costs casualties* and decides nothing.[26] [emphasis mine]

Tactical instructions issued to General Zach in 1800 echoed this sentiment, indicating that nothing really changed from 1796 to 1799.[27]

Higher command levels of the Austrian army were still organized along the 18th-century lines and strategic direction provided by the *Hofkriegsrat*, which was dominated by Baron Thugut. The *Hofkriegsrat* often caused more disunity in the direction of military affairs than its intended purpose of unification of strategy. Orders sent from Vienna were often irrelevant by the time they reached the commander in the field.[28] But this problem was not unique to the Austrians; Napoleon as emperor would have the same problem with his directions to his generals in Spain. The rather large (for that period) Austrian army was extremely expensive to maintain—45 percent of all the Imperial revenue went to its upkeep.[29] Austria's expensive army was essential to her existence as a dynastic state, and therefore was a more intrinsically valuable army from the standpoint of its rulers. This was one of two fundamental differences between the Russian and Austrian armies. High casualties were anathema to Thugut and the Holy Roman emperor. The Russian army, on the other hand, was a tool to expand Russia's influence in Europe and Tsar Paul's legacy of glory.[30] In the Russian army, casualties were often not a consideration, although Suvorov was more careful of his men's lives than most of his Russian contemporaries—but ruthless when he sensed victory.

The second fundamental difference between the two armies is related to the first. Russian operations and tactics took advantage of the ability of its soldiers to endure circumstances that the soldiers of other armies could not—in a word, the Russians, and especially Suvorov, molded their maneuvers around the operational durability of the troops. The Russian soldier often executed orders that no other troops of Europe could or would.[31] Russian units could endure terrific casualties and still remain cohesive. When one considers these attributes of the Russian army and of their leader, one finds that the French army's normally asymmetric advantage as a national and durable force composed of highly maneuverable units would count for little in the campaign of 1799 in Italy. Also, Schneider's requirement for real operational art requires this sort of symmetry—the "distributed" opponents operating in a similar fashion.

The overall Allied strategic plan was to attack France on all fronts, including opening an additional front in Holland by a combined Anglo-Russian force. Initially, the Russians were to have been employed as a component in all theater armies (except Egypt)—Holland, Italy, Switzerland, and Germany. As the various Russian columns made their way through Poland and then into central Europe, however, the French

Directory decided to go on the offensive in all theaters. These offensives had the purpose of overall defense, not conquest, essentially as spoiling attacks against the gathering forces of Austria, Russia, and Great Britain. The Directory, despite being spread thin on all fronts, instructed its main field commanders to take the offensive against the superior forces facing them in Italy, Switzerland, and Germany. This strategy was in part meant to hit and throw the Austrians back before the many Russian contingents arrived to reinforce them.[32]

The key area for all the belligerents, as it turned out, would be Switzerland (see map 7), whose terrain controlled the communications for both sides between the major theaters of operation in Italy and Germany. Additionally, whoever held Switzerland could threaten the flanks of their opponent or guard their own flanks for advances in adjacent theaters. The French placed arguably their best available general in command of the Helvetian (Swiss) army, Bonaparte's most reliable subordinate from his first campaign, General André Masséna. General Scherer, leaving his job as war minister for the Directory, was to have the effort in Italy once again with instructions to go on the offensive against the Austrian forces in northeastern Italy where the Adige served as the frontier between their forces and the French. Because events in Italy would be greatly influenced by Masséna's fortunes in Switzerland, it is to there we must first go before examining the campaign in Italy.[33]

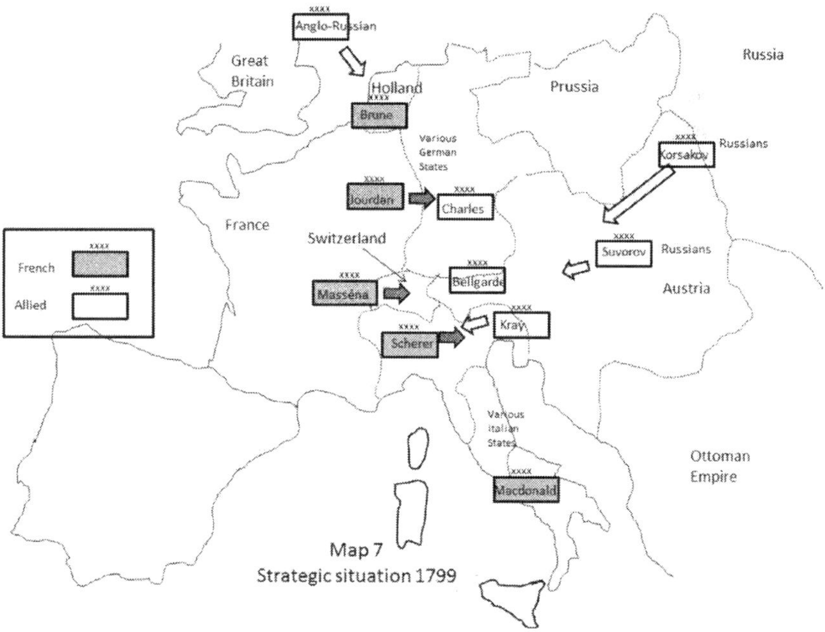

Map 7
Strategic situation 1799

PHASE I: SUCCESS IN SWITZERLAND AND DEFEAT IN ITALY

During the brief "interwar" period, the French had done at least one thing right in establishing a Swiss Helvetian Republic and thus gaining many of the hardy Swiss as allies as well as the right to station French troops at most of the key passes in western and central Switzerland. Masséna, the expert at alpine warfare, instinctively knew that his best course of action was to strike where the Austrians were most vulnerable, toward the passes in the southeastern portion of Switzerland known as the Grisons, here he could threaten or even invade the Tyrol, and thus threaten Austria's communications with both Italy and Germany via the Valtelline Pass (see map 8). Masséna opened his campaign in Switzerland in early March, commanding a force of around 50,000 troops (with 10,000 detached to guard the key St. Gotthard Pass). After additional deductions for key garrisons, passes, and communication, he had approximately 26,000 troops available for field operations. He initiated his offensive a week after the French armies in Germany had opened their campaigns and Masséna clearly caught the Austrians in the Grisons unprepared, achieving tactical and operational surprise.[34]

Masséna achieved success immediately, driving on Chur, the provincial capital, where he captured 3,000 prisoners on March 7. By March 26, he had entered the Tyrol; however, at this point the failure of Jourdan's army in Germany undermined Masséna's astonishing successes. Archduke Charles defeated the badly outnumbered Jourdan at Stockach on March 25 in southern Germany and then drove him back over the Rhine.[35] Now Masséna faced the might of Archduke Charles to his north plus the Austrian reserves in the Tyrol under General Bellegarde to his east and south.

Masséna gained command of what was left of Jourdan's army and now controlled over 70,000 troops, but he was defending from Mainz (Mayence) on the Rhine all the way to Scherer's area of responsibility in northern Italy. He withdrew from the trap that the western Tyrol had now become, although he retained control over most of the Grisons. On the Rhine he directed his subordinates to conduct an active defense along the Rhine (he remained with the Helvetian army in Switzerland). Austrian pressure eventually forced him to abandon most of the upper Rhine Valley and pull back his main force in Switzerland to the vicinity of Zurich by May 20. Throughout he fought numerous delaying actions as well as battling against Swiss Catholic peasants whom the Austrians had stirred up against his lines of communications. It is at this point we must return to the activities of General Scherer, upon whom the Hungarian general Paul Kray visited a series of defeats without Suvorov or his Russians present. These events forced Masséna to send precious reserves south to shore up the southern front and resulted in his eventual abandonment of Zurich (see map 8).[36]

The campaign began prior to Suvorov's assumption of command with French offensives along the upper Rhine in Germany and along the Adige in Italy. The

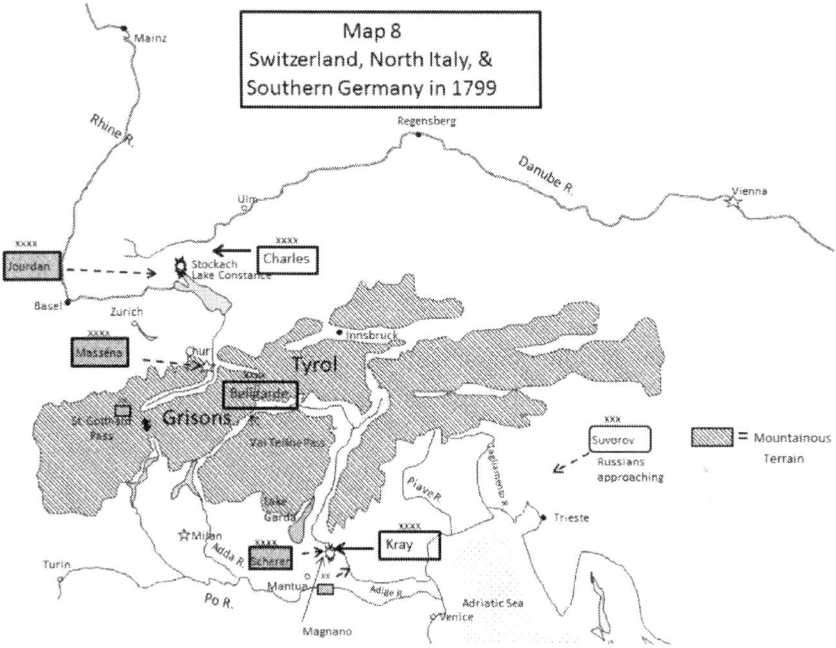

opposing forces in the Italian theater consisted of approximately 120,000 French versus 60,000 Austrians, but a large Russian column under General Rosenberg was rapidly approaching. The French were divided into numerous garrisons (due to simmering rebellion throughout Italy) and two field armies: the Army of Italy under General Barthelmy Scherer in the north (the same general whom Bonaparte had replaced in 1796) and the Army of Naples under General Macdonald in the south. These forces were spread across the length of Italy from the Alps to Naples, resulting in a local superiority for the Austrians along the line of the Adige River.[37] The Army of Italy was not the same fearsome force it had been under Bonaparte, however. Upon arriving in Italy on March 21 Scherer found it "indisciplined and disgruntled." Nevertheless, in accordance with the Directory's overall strategy, Scherer took the offensive in late March against the Austrian lines. Kray conducted a skillful defense à la Bonaparte along the river line, although he outnumbered Scherer. At first Scherer achieved some success when General Jean Victor Moreau, his second in command, led the left wing against the Austrians near the key terrain south of Rivoli and evicted them. This success did not last long. At the same time, Scherer fixed the Austrians in the fortress of Verona (their fortified bridgehead over the Adige) with the main effort under his personal command while sending a division on a wide flanking movement south of the old Arcola battlefield at Legnano as a diversion. His intent was evidently to pull the Austrians away from Verona and then assault or bypass it and win a battle on Kray's flank while he was busy to the south. The plan succeeded too well and Kray's Austrians savaged the division at Legnano, which included the Polish Legion composed of Austrian Polish deserters from Galicia (sometimes known as the Legion of the North). The Austrians attacked with particular fury against the Poles, whom they regarded as traitors, and drove this division back with great loss. The Poles, always good fighters, would be ill-used in this campaign as they fought against their hereditary foes who had partitioned their homeland—the Austrians and Russians.[38]

Scherer's column, meanwhile, made great efforts against the Austrians around Verona as Kray's attention was absorbed in the south, battling for over 18 hours against the local Austrian force of two divisions. Kray, once he realized the deception, moved rapidly through Arcola by a night march in a manner worthy of Bonaparte, and on March 27 he arrived just in time to turn the tables on the French. After two days of confused combat around Verona, the French asked for, and the Austrians granted, a one day truce—Kray well knew that a truce would only work in his favor since the Russians and Suvorov were marching hard to join him. The Austrians had almost 7,000 casualties and the French over 5,000, not including their losses at Legnano. On March 30, Scherer threw Serurier's division against Kray with disastrous results. The assault failed with over 1,500 French lost to that of barely 400 Austrians. Scherer's offensive only demoralized his army and cost him casualties he could ill afford.[39]

Moreau was called in to help Scherer, and predictably the rearguard forces at Rivoli were thrown back. On April 5, Scherer advanced again from his main camp at Magnano against Kray's forces just west and south of Verona for one more try to break the Hungarian general's resolve. Even with Moreau and extra reserves brought down from Rivoli, the Austrians had the slight edge in numbers (44,000 compared to the French 40,000) and were buoyed by their recent victories. This was one of the largest battles, both in geographic expanse and numbers of troops, to occur in Italy to this point in the wars and it was a severe French defeat. Kray, an expert in light infantry tactics, used a Suvorov-like tactic in his counterattack that decided the battle, an all-out grenadier attack on Scherer's right "with fixed bayonets" according to one observer. Kray's men inflicted over 8,000 casualties on the French, shattering the morale and confidence of Scherer's troops and their generals. The French fell back to the line of the Mincio River, anchored by the fortresses of Peschiera in the north and Mantua in the south (see chapter 2, map 4).[40]

The results of Magnano had serious consequences for the French. In addition to pulling back to the Mincio, Scherer recommended Macdonald and the Army of Naples come north, completely abandoning central and southern Italy; he also offered his resignation. At the same time, Scherer learned that General Heinrich Bellegarde, the commander of the Tyrolean army who now had troops to spare with Masséna in retreat, was sending strong flanking forces into his rear on the western side of Lake Garda, as Würmser had tried three years before against Bonaparte. Neither the army nor the general had the nerve for a Castiglione-type maneuver to retrieve the situation and the French were forced back to the Adda River by April 14, while the fortresses of Mantua and Peschiera were placed under siege. These garrisons subtracted another 7,500 troops from the French army, leaving it with less than 26,000 available troops. The Directory accepted Scherer's resignation and replaced him with General Moreau. The Austrians had good reason to be pleased. These victories, achieved without the assistance of the Russians and Suvorov, had given them a well-deserved pride in their own methods and abilities.[41]

On April 17, 1799, Suvorov assumed command of the Austrian force, arriving ahead of his own troops, who were at this point marching up to 30 miles a day. A point worth remembering was that the Austrians requested that Suvorov assume the overall command in Italy, even though the Austrians had the preponderance of force. Modern doctrine would have given an Austrian the command.[42] Actually, the first choice had not been Suvorov, but a variety of factors, including the untimely death of Frederick of Orange (a Dutch prince originally slated for the overall command) conspired to give the command to Suvorov. He was chosen by default due to his stellar reputation from the Turkish and Polish wars. Suvorov had already done two things to offend his Austrian partners. While in Vienna the old Marshal had refused to discuss his strategic plans with the members of *Hofkriegsrat*, including Thugut. Suvorov remained uncommunicative on his way to the front with his Austrian chief of staff, General Jaques Gabriel Chasteler, who also attempted to discuss operations with him.[43] Even so, when Chasteler arrived ahead of Suvorov at Kray's Headquarters, he was still bubbling with enthusiasm about his new chief. Chasteler was a figure not unlike Berthier, a professional staff officer of considerable talent, and Suvorov gave him carte blanche in reorganizing all of the Russian and Austrian units "along the most modern lines, transforming [the army] from a collection of individual regiments into permanent multi-regimental all-arms divisions." Chasteler gave each Austro-Russian division its own staff, composed of trained Austrian staff officers, with one battery each of horse artillery (which was still a recent innovation) and medium artillery, the latter composed of the excellent Austrian 12-pound guns plus one howitzer.[44]

Suvorov's second tactless act was to issue a series of instructions to the Austrian officers delineating which tactics to employ and under what specific circumstances. One of Suvorov's directives detailed that the infantry was to form into "two lines" from the column at one thousand yards from the enemy and then advance slowly to

> three hundred yards, this being the maximum range of the excellent French musketry. Then, when the order to advance was given, the troops should proceed at the usual rate,

> but after a hundred paces they must double their speed, and a hundred yards short of the French line they must rush at it . . . with bayonets leveled. The second line was *to follow through* while the first, having delivered their charge, halted and reformed.[45] [emphasis mine]

Here was the idea of the echeloned attack (albeit at the tactical level) where the momentum of the offensive could be maintained by a fresh second echelon while the first reconstituted, presumably to be passed through the second echelon at some later point if need be. Suvorov announced to his army that, "We have come to beat the godless, windbag Frenchies. They fight in columns and we will beat them in columns!"[46] There is an apparent contradiction here between the use of columns or lines that is easily explained. Suvorov used both. This again goes back to his transition between operational maneuver and grand tactical employment for battle, usually a meeting engagement. His purpose was to maintain the impetus of his attack on a broad front while maintaining the advantage in movement and cohesion provided by the column. As we shall see, this operational method was to work quite well with the Austrians, in fact they may have executed it more skillfully than their Russian comrades who often did not form line but remained in column for the duration of an attack.[47]

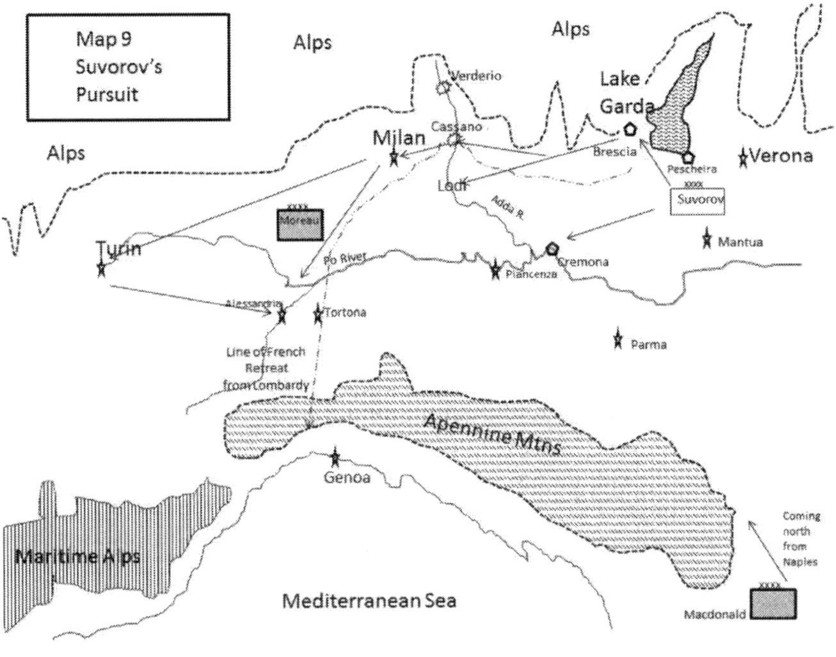

Suvorov conducted training courses to ensure the Austrians digested his instructions properly. This action could not help but have some negative consequences. The future Field Marshal Joseph Radetzky wrote of these instructions: "the order . . . was dictated to a victorious and confident army . . . The consequence was an extraordinary division between the allied forces—a division which extended all the way up to headquarters." Not all of the Austrian officers were so alienated and Radetzky, chief of staff to General Michael Frederich Melas, Suvorov's second in command and commander of the Austrian component of Suvorov's force, may have simply been reflecting the views of his chief.[48] Chasteler, in particular, made heroic efforts to work with Suvorov. He and the Cossack leader Denisov established a "joint" intelligence cell that combined Austrians, Cossacks, and Italians together to interpret the crude maps that the Cossacks made to describe their reconnaissance efforts.[49]

These affronts were initially mitigated by Suvorov's victorious advance across Italy (see map 9). Suvorov maintained generally good relations with his Austrian staff and wore an Austrian rather than a Russian uniform to emphasize his spirit of cooperation.[50] Suvorov's intent was to give the French no time to recover and to liberate Lombardy while preventing Macdonald from joining up with Scherer (still in command for a few more days). Leaving Kray to prosecute the sieges of Peschiera and Mantua, he advanced northwestward and along the right bank of the Po with about 50,000 Austrians and Russians. First to fall, with its entire garrison (1,130), was Brescia on April 21 after a brisk bombardment and a threat to put the entire garrison to the sword if they did not surrender. To the south, an Austrian detachment advanced on Cremona, which was seized on April 22.[51]

It was during this hectic time that Moreau replaced in command of the shrinking French army. At Lecco, General Serurier repulsed the northern arm of Suvorov's assault on the Adda; however, Moreau was defeated in a series of engagements around Cassano on April 26–27 as Suvorov forced the Adda River line, including, ironically, a barely contested crossing at Lodi, even though the French had destroyed the bridge there. Suvorov then turned his forces north against Serurier, who had not fallen back after the defeats around Cassano. At a bend in the Adda River near Verderio he was surrounded and had to surrender himself and most of his division (now down to about 3,000 troops). Suvorov's energy and use of multiple lines of advance for his divisions had accomplished the first goal Emperor Francis had set for him—the arrival of the joint army on the western side of the Adda.[52]

Suvorov then set out, in disobedience of his orders, to capture Milan, capital of Lombardy. During this advance the Austrians seem to have executed Suvorov's guidance faithfully. At St. Juliano a joint Austro-Russian division executed the double-line echelon tactic in an attack that blunted a counterattack by Moreau.[53] Milan fell on April 29 and Suvorov forced Moreau into defensive positions in the mountains around Genoa, almost precisely the same location that Bonaparte had started from three years earlier. Suvorov's offensive equaled that of the Corsican in energy, brilliance, and execution. Northern Italy now lay uncovered and the Allies settled into a consolidation of their gains by placing all of the major fortresses with French garrisons under siege. Suvorov additionally invaded Piedmont, reaching Turin on May 27 and putting that place, Tortona, and Alessandria all under siege.[54] As usual, the French left too much of their combat power cut off and besieged. However, Suvorov would soon find his operations challenged as a new French army composed of undefeated veterans under a determined general rushed north to confront him.

OPERATIONS AROUND ZURICH AND THE TREBBIA

Suvorov's operations had a profound effect on those of Masséna. Until this time Masséna had been holding his own, but the collapse of the French forces guarding his southern flank meant that he had to pull back. Initially, he had hoped to stabilize his front from Zurich to the left bank of the Rhine to account for the defeat of the French Forces in Germany. Because of the collapse of Scherer's forces and the subsequent loss of Lombardy, Masséna received orders to send over 15,000 reinforcements south to help stabilize the situation on May 6. Nonetheless, he attempted to hold his position around Zurich.[55]

Archduke Charles gave Masséna no chance to rest. Advancing from the east he assaulted the French positions in front of Zurich on June 3 and was bloodily repulsed. When he returned on June 4 for another round of assaults, Masséna again skillfully drove the Austrians back, but his own Helvetian army had also suffered grievously. Rather than lose his entire army to a third Austrian assault, Masséna conducted a skillful retreat to a more defensible terrain to his west along the Limmat River and the Albis Mountains (see map 10). Masséna's defensive victory at Zurich in June 1799 reminds one of the Duke of Wellington's victory in Portugal at Busaco in 1810 against Masséna himself (see chapter 7). In both cases, a fine defensive victory was followed by an operational retreat to a more advantageous terrain—a superb example of defensive, attritional warfare in the hopes of one's enemy making a mistake and creating the opportunity for a counterstroke. Masséna's new position also provided better protection for his southern flank, which had been uncovered by the conquest of Lombardy and the threat now posed by Suvorov to Piedmont. At the same time, Masséna's army was renamed the Army of the Danube and his command limited to Switzerland as a new Army of the Rhine was formed from the forces holding along that river north of Basel. Masséna also received reinforcements directly from France equating to a division of infantry and a brigade of cavalry.[56]

The Directory, hoping to retrieve the situation in Italy and relieve the besieged fortresses, recalled General Macdonald and his Army of Naples from southern Italy

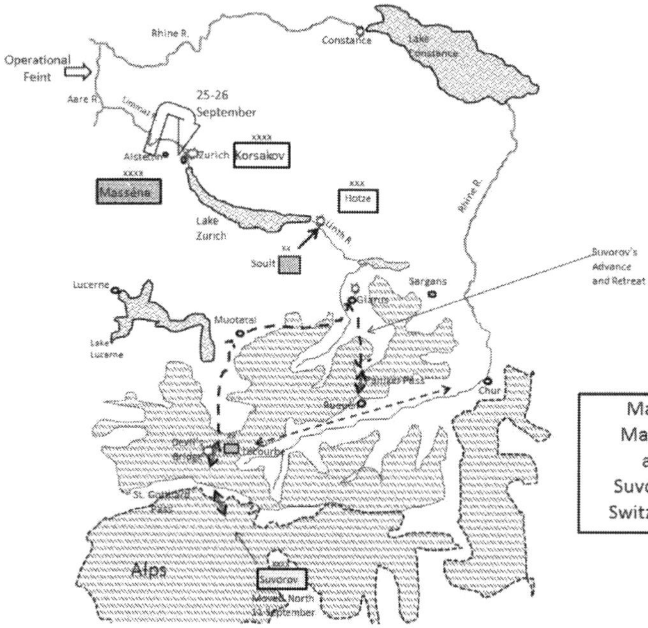

Map 10
Masséna
and
Suvorov in
Switzerland

in order to unite with the remnants of Moreau's army. Suvorov had, to a degree, anticipated this move and his requests for reinforcements brought 11,000 more Russians detached from Rimsky-Korsakov's corps (on its way to Switzerland) and another 10,000 Austrians under Bellegarde from Charles's army in Switzerland. Suvorov also suspended the siege of Mantua and detached Kray, the victor of Magnano, to the south to delay and develop Macdonald's force as it advanced rapidly through Tuscany. Despite these Russian precautions, Macdonald and Moreau almost succeeded in uniting. Macdonald's unexpectedly rapid advance to the Trebbia River took even Suvorov by surprise. Suvorov was not used to his enemies advancing as rapidly as he! Moreau's job was to bring his now reinforced army into a position on Suvorov's right flank where he could either threaten Suvorov or unite with Macdonald, or both.

While Suvorov rushed with all available troops toward Tuscany (marching almost as far and certainly as fast as the French), Macdonald resumed his advance with an army of approximately 33,000 veteran troops. On the June 11, 1799, Macdonald collided with the advanced units of Austrians (General Ott's division) at Modena (see map 11). The Austrians again used Suvorov's tactics to good effect, this time bayonets against cavalry, and repulsed the French, although they lost part of their rearguard as prisoners to the French.[57] As one reads these accounts of *Russian* tactics and bayonet charges one almost forgets that the troops doing the executing were *Austrian*. Macdonald advanced again on June 12 with more success and forced the Austrians in the center to retire on Mantua. Macdonald now turned to the northwest and the expected union with Moreau. The only force standing between the two French generals were the 6,000 Austrians under General Ott—but Suvorov, with 22,000 troops, was rushing to Ott's aid. Additional Austrian forces under Melas were following in Suvorov's wake. Macdonald had already pushed the retreating Ott over the Trebbia and halted between that river and the smaller Tidone River. The Trebbia was to be an engagement that exactly conformed to Suvorov's style, a meeting engagement that grew over a number of days as new echelons arrived in the hope of shattering the French army and driving it away from Moreau.

The battle of the Trebbia began on June 17 as a skirmish by one of Macdonald's divisions (Rusca's) as it pushed across the Tidone. Ott's Austrians were slowly withdrawing under this pressure when Suvorov and his advance guard, mostly Cossacks, arrived and counterattacked. The French, who had not expected a battle, were thrown back in disorder to the Trebbia with some units remaining on the western side. Macdonald claimed he had specifically ordered an avoidance of a battle until Moreau was in communication.[58] The battle of the Trebbia, which started out as

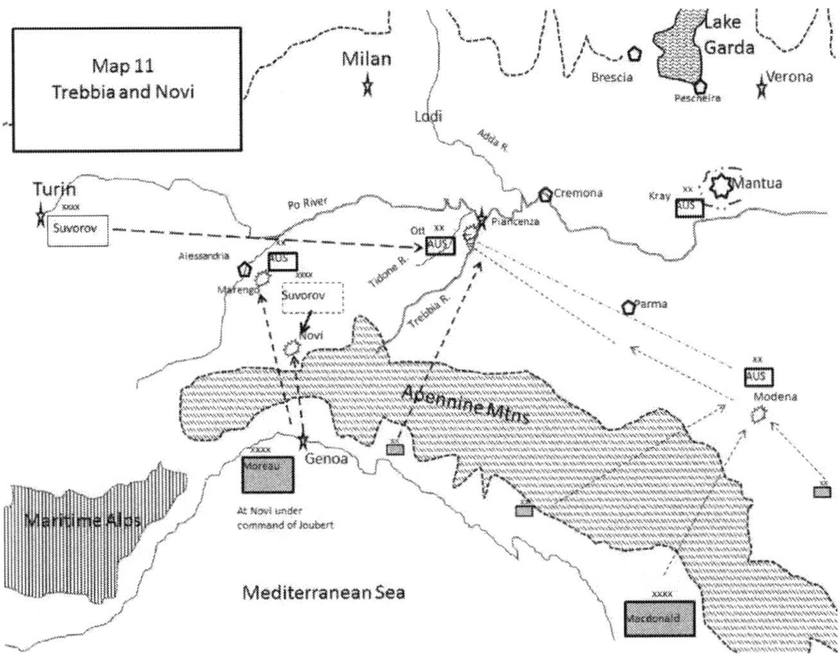

a meeting engagement between an advance guard and a rearguard, turned into a three-day battle of wills between Suvorov and Macdonald. In this it reminds one greatly of the Battle of Gettysburg during the American Civil War.

Macdonald resolved to remain on the defensive to await the arrival of the rest of his forces. Suvorov, on the other hand, formed his army for the attack during the night as his tired troops marched up to the battlefield. Suvorov was not able to organize his forces and bring them up until the next afternoon. In doing so, he achieved a certain amount of surprise against the French on the left bank of the river. Macdonald withdrew Salm's French division in haste and with loss to the right bank. The battle then raged into the night, the cannon only ceasing their racket at about 11 o'clock. Macdonald described the Austro-Russian attack as a "vigorous onslaught . . . their strength was great, and their cries and howls would have sufficed to terrify any troops except French ones."[59]

Suvorov's approach on the second day had been to advance in five large columns, which he again directed to "form line immediately in an orderly fashion . . . without *pedantry* or . . . excessive . . . *exactitude* [emphasis mine]." Macdonald counterattacked each of these columns in turn and nearly succeeded in forcing the Russians back. The Russian soldiers' refusal to break after these punishing blows sustained Suvorov's position on the west bank of the Trebbia as night and exhaustion brought the second day's battle to a close. Suvorov's decision to attack with his outnumbered and exhausted army had paid off—for the moment.[60]

The next day, both commanders decided to attack. Macdonald temporarily had the advantage in numbers and concentrated 12,000 of his troops for a decisive flank attack against the Russians under Peter Bagration on Suvorov's extreme right. Macdonald supported the main effort by another turning movement on the left. These troops included the Polish Legion under Dombrowski, which had fought furiously against Suvorov the previous day. A Russian counterattack practically annihilated the Poles, but this attack in turn opened a gap around Suvorov's central division under Schveikovsky. In turn, this gap permitted an opportune counterstroke by Generals Victor and Rusca that almost broke the back of the Russian position. The crisis of that day's battle had arrived. Both Bagration and Rosenberg requested to withdraw. At this point, Suvorov became personally involved in animating his troops. Macdonald heard that Suvorov went so far as to threaten to kill himself if his troops withdrew another step. Melas, the senior Austrian general present, exacerbated the crisis by bringing only part of the reserve into the bloody melee in the center. However, the Russians held and the third day's battle died down with both sides still separated by the river. Macdonald and his surviving subordinates had had enough. After a council of war they withdrew under the cover of darkness with their campfires still burning and retreated through the mountains to the west, toward Genoa. When Macdonald arrived there, he and Victor were summoned to Paris to explain themselves.[61]

When the Cossacks advanced early the next morning they found the French gone and Suvorov ordered an immediate pursuit. What little pursuit he was able to mount managed to capture the bulk of Macdonald's rearguard, thus demonstrating how even the most meager pursuit by reserves can deliver great rewards. Macdonald acknowledged that he had received "disastrous . . . losses" and estimated the Allied casualties as equally "enormous." The butcher's bill was sobering: including the rearguard actions, Macdonald had lost half his army, some 16,000 killed, wounded, and prisoners. The Allied losses, for the size of their army were nearly as bad, 6,000 killed and wounded Russians and approximately 4,000 to 5,000 Austrians. Meanwhile, Moreau's advance near Marengo had cost the Austro-Russians another 2,000 casualties due to an abortive counterattack by General Bellegarde. It is clear that Moreau's sluggishness in coming to MacDonald's aid, despite his success near Marengo, probably cost the French the battle.[62]

Operationally, the Battle of the Trebbia highlighted for both sides the importance of rapid maneuver over long distance and then fixing one's enemy by feeding reserves into the battle as they arrived. Suvorov's goal was always to prevent a juncture of the two French armies and his skill in doing so reminds one of Bonaparte's actions at Castiglione and around Mantua. The French lost at least twice as many casualties (not including their rearguard) as the Austro-Russian army, a clear testament to the training and combat power of both the Russians and Austrians against a veteran, well-led French army.

A final note on the battle of the Trebbia concerns the actions of General Melas. The Austrians had already been campaigning longer and, to a certain extent, were fatigued. But the Russians were also spent after their numerous forced marches. One historian has noted that it was "curious" how slow and ineffective Melas was in supporting Suvorov on both June 18 and 19. All accounts of the battle agree that Melas responded slowly. Even Macdonald admits that had he been hard- pressed on the second day (June 18) he did not know what he should have done.

Rather than improve the unity of effort of the coalition, the conduct and results of this bloody battle combined with other negative trends to undermine the unity of the coalition in this theater. In May, prior to the Trebbia battle, some of the Austrian generals "complained about Suvorov's *crude tactics and unnecessary losses* and risks (emphasis mine)."[63] Suvorov had already lost his excellent chief of staff Chasteler due to Austrians who schemed against that officer simply because of his devotion to Suvorov. Colonel Weyrother, Chasteler's replacement, became just as devoted to Suvorov, but this move indicates trouble in the coalition ranks was starting to have a negative influence.[64] Was Melas in fact one of the officers that had been complaining about casualty rates and did he rectify the situation locally during the battle by dragging his feet? No solid proof exists, but the "curious" nature of his actions argues that he may have.[65] Up to this point, Thugut had generally backed Suvorov against complaints by

Austrian generals, despite the fact that Suvorov's high casualties, particularly among his Austrians, ran counter to the "first priority" of the Habsburg dynasty—the preservation of its army. Suvorov had already alarmed Thugut by calling on the Piedmontese to rise up and support their exiled king, which was contrary to current Austrian policy of absorbing Piedmont in some fashion after the successful conclusion of the campaign. Immediately after the battle Suvorov received word that Turin (Piedmont's capital) had surrendered. Suvorov, following the Tsar's orders, used this event as an opportunity to summon the Piedmontese king from Sardinia to resume the throne. Thugut immediately disavowed Suvorov's actions and subordinated him to his personal control.

Suvorov's Austrian subordinates now began to speak openly against him. Thugut no longer discouraged their bypassing Suvorov to communicate directly with him. Additionally, Suvorov was now required to report to Vienna prior to engaging in battle. Prior to leaving for Italy Suvorov had written the Tsar a confidential note stating his intention not to "waste time in sieges." The simmering debate over strategy and the operational means to inform it—sieges versus field engagements—was now resolved in Austria's favor. The Austrian strategy involved a reduction of the remaining fortresses in Italy before resuming field operations against the French. The losses during the Trebbia fight had gutted Suvorov's Russian contingent and he was now very dependent on the Austrians to conduct the sieges and for replacements. In this manner, not out of deference to Thugut but because his control of resources was limited to his Russians, Suvorov had to submit or resign. Paul rejected any suggestion that Suvorov be replaced. These internecine squabbles, as already discussed, became a major factor in all parties desiring to move Suvorov to another theater.[66]

The course of the campaign during July now assumed a slower pace as Suvorov bowed to the will of Thugut and the *Hofkriegsrat*. General Moreau was masked in Genoa while the sieges of Mantua, Tortona, and Alessandria continued. The French had other plans. They now sent General Joubert, one of the stars of Napoleon's first Italian campaign, to take command from Moreau, who was to move north to command the Army of the Rhine. Joubert's mission was a component of a broad scheme of French counteroffensives running from southern Germany to Italy. Joubert was reinforced to a total of approximately 40,000 troops, which gave him a field force of approximately 35,000 troops.[67] His goal was the relief of the beleaguered French garrison of Mantua. Joubert wisely retained Moreau for the time being as an adviser for the upcoming offensive.

Mantua, however, had surrendered on July 27 and Alessandria even earlier on July 21. These surrenders released considerable Austrian reserves for operational use, especially Kray's outsized division (almost a corps) from Mantua. Joubert was unaware of these facts as he advanced in four columns on August 11 toward the Po River valley from his bastions in the Maritime Alps. Suvorov, reinforced with 10,000 more

Russians (diverted from Korsakov), now rushed to engage Joubert with over 51,000 troops in hand. He also issued orders for his screen troops not to impede the French advance so that he could "lure the French down from the mountains to the plain and crush them."[68] Again, Suvorov's operational design was a meeting engagement with the unready French columns to be followed by echelon after echelon arriving over time until his enemy was overwhelmed. As usual his Russians and Cossacks would lead the way. On August 14, Joubert's divisions united at Novi and halted. Suvorov's army, with the exception of Melas's column, had also arrived before Novi. Joubert, surprised by the size of Suvorov's force, which was larger than his own, now contemplated retreat. While Joubert hesitated, Suvorov attacked.

Joubert's defensive position had its left flank firmly anchored in the Maritime Alps at Pasturana with strong defensive positions in the center around Novi. Joubert's right was weak and "in the air," but it seemed safe because Melas was still distant and Joubert was unaware of his approach. Suvorov's conduct of the battle was simple. He intended to fix the French with heavy attacks on their left, then their center. As these attacks proceeded, he would deliver a flanking blow using Melas against the French right. Finally, he used wider flanking movements, one composed of a cavalry force and another Russian contingent under Rosenberg, to maneuver on the French rear. The Austrian general Kray "begged" Suvorov for the honor of opening the attack on the French left with his corps. Of note, Suvorov was routinely organizing his army into multidivision corps, sometimes even combining Austrian forces with Russians—predating to some extent the formalization of the corps as an operational organization by the French under Napoleon.

The attack began at daybreak, the first of 10 attacks by Kray that day.[69] It caught Joubert off-guard and he was killed as he reconnoitered forward of Pasturana with his skirmishers. Moreau, confirming Joubert's wisdom in retaining him, took charge and another contest of wills began as Suvorov attempted to pry Moreau out of his defensive positions. Success depended on Melas, who had been so slow at the Trebbia. He did not disappoint his Russian commander on this day. The battle was nine hours old when Melas arrived and opened his attack on the French right, which Moreau had weakened to support the vicious attacks on his left and center. Melas's actions, along with the deep penetration near the French rear by the other two columns, convinced Moreau that his position was untenable and he attempted to withdraw. Moreau's artillery became entangled in the narrow village streets and mountain roads, blocking and disordering the retreat. Many of his troops were taken prisoner.[70]

The price paid for this success was severe losses incurred by Kray's Austrians and Bagration's Russians. The final result was nearly 12,000 French casualties (4,000 of them prisoners) as opposed to almost 8,000 for the allies—the greater percentage of them Austrian. One historian has claimed that "Even at the Trebbia the fighting had not been so inhumanly bitter and obstinate as it was here."[71] The last French field force in Italy had been defeated and would soon be confined within Genoa for an epic siege. It was ironic that the battle of Novi, where cooperation between Russians and Austrians on the field of battle was never exceeded again in that generation, stood in stark contrast to the squabbling, distrust, and outright enmity that existed between the governments represented by Suvorov's army. However, the cost in Austrian lives was enormous by 18th-century standards.

The fragile unity of effort that still existed in the field was soon to collapse at higher levels due to three factors: (1) the strategic shift of Suvorov to the north, (2) the dispute over the issue of sieges versus an annihilating pursuit and expulsion of the French from Italy, and (3) the cumulative heavy losses to the Austrian army in all theaters, but especially Italy. A week after Novi, Suvorov finally received official orders for his transfer, with his Russian troops, to Switzerland. He was opposed to this move because "with the loss of Italy, one cannot win Switzerland." However, Suvorov's disputes with the *Hofkriegsrat*, which had ordered him to cease his pursuit of Moreau, finally convinced him to observe, "I cannot continue to serve here."[72] Suvorov's problems with his subordinates only worsened. Melas resumed his hostile attitude and refused to obey Suvorov's orders if they did not correspond with the orders he now received from Vienna. Although Suvorov had won the decisive battle of the Italian campaign, the Austrians no longer felt the need for his services in Italy. Baron Thugut's correspondence reveals the prevailing Austrian attitude toward Suvorov at this juncture:

> The innumerable inconveniences caused in Italy by the conduct of *this general* [italics mine], acting according to foreign [Russian] orders and obviously opposed to the interest of His Majesty, are of such a grave nature that His Majesty could easily do without his commanding an army [in Italy] if having him sent to Switzerland is happily followed.[73]

The Austrian generals were no longer answerable to Suvorov, unless, like Kray and Chasteler, they risked their standing with Vienna by cooperating with him. This should have been no surprise to the old field marshal. Under Austrian officers the casualties would certainly have been less, although the results perhaps less decisive.[74]

MASSÉNA AND SUVOROV IN SWITZERLAND

It is fitting that the closing chapter of the War of the Second Coalition in 1799 should bring into conflict the two operational artists we have been following: Masséna and Suvorov. Before bringing them together, though, we must return to the period prior to the Battle of Novi, as Masséna slowly improved his operational position in Switzerland and then created his own masterpiece around Zurich. Recall that by late June

Masséna had stabilized his position east of Zurich along the Aare and Limmat Rivers. In the southern portion of his theater, an independent division of 12,000 troops under General Jacques Lecourbe operated east of Lucerne (see map 10). Masséna made successful probing attacks to the east at Alstettin and Lecourbe probed toward the critical St. Gotthard Pass by which Suvorov would have to come north with his Russians once he left Italy after the battle of Novi.[75]

By August, Masséna was ready for more than tentative measures and had Lecourbe move directly against the St. Gotthard Pass, guarded by one Austrian brigade. Lecourbe, as much an expert in mountain warfare as Masséna, seized the critical pass on August 18. Masséna had his forces, principally the divisions of Generals Soult and Lorges, demonstrate on the Limmat to distract Charles.[76] Charles naturally assumed that offensives before Zurich and in the south must mean the French left was weak and he attempted to throw a bridge across the Aare River on August 17, but this attempt was handily defeated by light artillery and Swiss riflemen from Masséna's reserves. Finally, other French forces further secured Masséna's right flank against the Austrians by capturing the town of Glarus to his southeast on August 29.[77]

It was at this point—with Masséna reinforced to a strength of over 87,000 men, a field force of about 40,000 under his direct command west of Zurich, and much good ground in his possession for further offensive operations—that the Allies ordered Charles north for operations in Germany and Suvorov north to combine with General Rimsky-Korsakov's newly arrived corps in Switzerland. Charles received his order to depart with the bulk of his Austrians the day after his repulse on the Aare River (August 17) and Suvorov received his own orders to depart Italy with his Russians a week after Novi. Charles had protested this plan but duly executed his orders due to his new focus for operations being much further north along the Rhine near Mannheim. Masséna was presented with a golden opportunity to retake Zurich and destroy Korsakov's army before Suvorov or Charles could come to his aid.[78]

As August came to an end, Masséna found to his front approximately 48,000 Russians and Austrians with Korsakov covering the front from the Rhine to Zurich and about 24,000 Austrians under General Hotze covering the northern shore of Lake Zurich and the Linth River.[79] It was against this extended front that Masséna now massed. To the south, Lecourbe was directed to continue his own offensive toward the Grisons against the strung out forces of Hotze, leaving only light forces covering the St. Gotthard Pass. Suvorov approached this pass with over 21,000 of his Russians and an Austrian brigade under Auffenberg assigned to join his command once he forced the St. Gotthard Pass.[80]

Masséna had all the advantages of interior operational lines during this delicate period, but he had to act before Suvorov erupted from the St. Gotthard Pass on his flank and into Lecourbe's rear. General Bernadotte, the new war minister for the Directory, informed Masséna of this opportunity on September 5 after Masséna reported Charles's departure. By September 20, Masséna had resolved to attack in three different places: with his main army against Korsakov at Zurich, using the division of Soult against Hotze on the Linth, and with Lecourbe, as already mentioned, toward the Grisons. His main effort at Zurich involved an operational deception against the widely deployed Russians on their right flank near the Rhine, to distract their attention, while he sent a column of 14,000 men over the Limmat toward Zurich. At Zurich he employed tactical deception by stealthily moving cannon and bridging elements directly opposite the city. The second Battle of Zurich began on September 25 with these cannon waking up Korsakov and his troops in the city, achieving both operational and tactical surprise. The crossing of the Limmat occurred without opposition as Korsakov focused on the forces immediately to his front. The key to the battle was the seizure of the Zurichberg high ground and this Masséna's main column did at great cost while at the same time it seized the western suburbs. By the end of the day the French had trapped Korsakov inside the city, where Masséna offered the Russian the opportunity to evacuate the town without further bloodshed. Meanwhile, to the south, Soult had also achieved tactical surprise in crossing the Linth and eliminating any possibility of Hotze's Austrians coming to Korsakov's aid. The brave Hotze was killed at the beginning of this engagement.[81]

Korsakov, learning of Hotze's death and Soult's success, refused Masséna's offer and resolved that night to retake both the suburbs and the Zurichberg while he broke out with the rest of his army to the north (probably to save his many cannon). These attacks were initially successful, gaining Korsakov some time, but ultimately they resulted in many more losses as the French counterattacked and slaughtered the packed Russians as they retreated through the streets of the city. Korsakov had lost over one-third of his army, 9,000 troops, including over 5,000 prisoners. The losses of the Austrians to the south were also heavy. On October 7, Korsakov lost another 3,700 men in an ill-advised attempt to regain some of the initiative when he attacked the far left of the French line along the Rhine. After that his force was no longer capable of offensive operations. At the other end of the wavering Allied line in the north, held by the émigré corps of the Prince of Condé at Lake Constance, Masséna continued his operations, taking Constance on October 8.[82]

In the south, Masséna gave the Austrians no rest, attacking with Soult who managed to link up with Lecourbe, inflicting over 5,000 casualties and causing the leaderless Austrians to retreat. But now it was the French turn to go on the defensive as Suvorov's army entered the theater of operations, sweeping away the French forces at the St. Gotthard Pass and then driving hard for the next mountain defile at the famous "Devil's Bridge." Here, by sheer willpower and with much loss, Suvorov forced an almost impassable defensive position against Lecourbe's troops and artillery. As Lecourbe fell back he was joined in the nick of time by Soult coming down to reinforce him. Together these two divisions repulsed Suvorov's advance guard. It was at this time, on September 28 that Suvorov learned of Korsakov's disaster and realized his army was probably Masséna's next meal.[83]

Suvorov began to retreat toward Glarus to cross the Linth River. He did so knowing that he was essentially bottled up in an alpine pass with French and Swiss forces to his front and even larger French forces in his rear. Nonetheless, he battled his way through the forces between him and Glarus, arriving at that town early on October 1. For the rest of the day his advance guard fought a desperate action against superior French forces on the Linth north of Glarus, but the French repulsed a series of uncoordinated and furious Russian assaults across the river. The way north was seemingly barred by strong French forces. By October 4, his rear corps under Rosenberg had closed up after an exhausting march from Muotatal. For one of the few times in his career Suvorov made a poor decision. Influenced by the Tsar's relative Prince Constantine, who accompanied the army, Suvorov accepted a majority decision of a council of war on October 2–3 to reject another try (proposed by Weyrother) along the northern line of retreat toward the Austrians at Sargans. This was what Archduke Charles hoped he might do and he pushed more troops in that direction to join with Suvorov once he punched through the French. However, Suvorov decided to retreat over the Alps through the almost impassable Panixer Pass, which meant he would have to abandon his artillery since only pack animals could make it through that route. It is hard to say what might have happened had Suvorov followed the advice of Weyrother and Archduke Charles. Condé's defeat at Constance on October 8 might still have made the northern route untenable given the success of French arms. However, the Russian generals under his command no longer trusted the Austrians and the route to the south was chosen. Suvorov abandoned his cannon, wounded soldiers, and trains and set off.[84]

The southern route proved perilous enough. Suvorov was racing against time as Masséna had four French columns closing in on him: Gazan's division to his north, Molitor's division to his west, Loison's division to his south, and another wide flanking column of one battalion to his east. The most dangerous column was that of Loison, which if it could reach the southern end of the Panixer Pass would have an overwhelming advantage in numbers and artillery that Suvorov probably could not overcome as two more French divisions closed on his rear. Suvorov won this race, but at great cost as his men collapsed in the deep snows and high altitude and were left to die or be taken prisoner. On October 8, when his men crawled into Rueuen on the Vorderrhein River, the force of over 21,000 Russians numbered no more than 11,000 effectives, the Cossacks, amazingly, having fared best.[85]

Masséna and Suvorov had not actually fought within eyesight of each other in this final operational match, but that is the way of operational art, one's combinations are often unseen and frequently dependent on the initiative and leadership of operational subordinates. Masséna can be said to have defeated Suvorov at Zurich, even if he did not defeat him in person during the harrowing Russian operations in the heart of the Swiss Alps. Masséna went on to earn the sobriquet of "Dear Child of Victory" and the acclaim of the Republic while Suvorov received orders to return to Russia with his army and was dismissed in disgrace by an ungrateful Tsar. The old Russian died, heartbroken it is said, not long after in 1800.

* * *

In conclusion, the collapse of the Second Coalition was due to many factors, not the least of which were the divergent war aims of Austria, Russia, and Great Britain. Nonetheless, Suvorov's handling of the Austro-Russian army in 1799 in Italy strained an already fragile unity of effort. Suvorov saw the fine instrument that he had in the Austrian infantry and used it to maximum advantage. The Austrians performed, in many respects as well or better than their Russian comrades, whose very nature dictated the tactics that Suvorov employed. In the process Suvorov damaged this instrument with excessive casualties. More importantly, it was never "his instrument" to begin with, it was only on loan. He may have better served his cause to have understood the constraints for its use before decimating it in two bloody battles. The wastage of Austrian troops since the war's beginning in March, a scant five months earlier, now numbered more than 100,000. The Habsburg dynasty could ill afford to continue these bloody campaigns and hope to have an army to control their empire, much less fend off the Turks, Prussians, and even Russians.[86] The Austrian decision to let the Russians assume the main offensive burden at the end of the summer of 1799—in the light of all this evidence—is not surprising. It was not so much the French who defeated Suvorov as it was the disunity and strategic combinations of the political leaders of the Second Coalition. In the final analysis, Suvorov practiced operational art to a high degree in Italy and Switzerland, but he did not practice it successfully within the framework of coalition warfare. Thomas Graham's contemporary comment at the end of his narrative of this campaign sums it up: "[Suvorov] was the best possible General for the *Russians* [emphasis mine]."[87]

FOUR

ARTISTS OF THE WAVES

THE ROYAL NAVY, NELSON, AND OPERATIONAL ART AT SEA[1]

This chapter looks at the commander of the British fleet at Trafalgar, Admiral Lord Horatio Nelson, baron of the Nile, and the campaign that led him to engage the French and Spanish combined fleet off Cape Trafalgar on October 21, 1805. As battles go, Trafalgar numbers among the most decisive in modern history and remains a monument to the superior training, tactics, organization, and—especially—leadership of the Royal Navy of the period. Nelson left a profound legacy to command of the sea that still affects how navies, and especially the United States Navy, operate today.[2] Nelson's career, both culminating and ending at Trafalgar, highlights basic truths of naval operations, truths that endure to this day.

Nelson's "touch" codified for the Royal Navy and all other navies the guiding principle of centralized operational planning and decentralized, violent execution in combat. All of Nelson's operations and battles highlight these simple principles, but none more so than his operational art masterpiece off Cape Trafalgar over two centuries ago. The campaigns of the Royal Navy, and especially of Nelson, leading to Trafalgar also reflect the role that naval strategy plays in war, especially between two empires whose sources of power differed so strikingly that the contest was supremely an asymmetric one—but this asymmetric character did not fully manifest itself until *after* French and Spanish sea power had been smashed over a period of 11 years, at which point it never recovered, despite the best efforts of Napoleon Bonaparte.

Ironically, the battle was not decisive in the ways some have characterized it—neither for its immediate impact on the general war in Europe in 1805 nor for its impact over the long term on strategic events.[3] Simply, the Battle of Trafalgar was decisive because it both embodied the successful application of superior sea power by the major European maritime power and emphasized the *necessary* role that command of the sea by Great Britain would play in shaping the eventual defeat of Napoleon.

In and of itself the achievement of what Sir Julian Corbett referred to as *general command of the sea* through the mechanism of a decisive battle at Trafalgar was not *sufficient* to guarantee the British victory in their long struggle with France.[4] Great Britain's eventual victory relied on many other factors and agencies including the actions of Napoleon himself. However, command of the sea, based upon the Royal Navy's blockade and its victories like Trafalgar, was a *necessary* condition without which Napoleon might not have been so frustrated that he felt compelled to counter British sea power with an ultimately self-defeating economic blockade—the Continental System.

In another sense, the road (or rather the cruise) to Trafalgar is far more interesting for the insights and explanations it offers on institutional culture, the importance of training and doctrine, and most importantly the British legacy of naval leadership and operational command and control at sea. The essence of this legacy encompassed *centralized planning* and *decentralized execution* and became the basis for modern U.S. naval doctrine.[5] It will also highlight a command philosophy the U.S. Army of the present era labels *mission command*.[6] This chapter will examine how Nelson prospered under this system at the Battle of Cape St. Vincent, executed the system as an operational commander in the campaign that led to his overwhelming victory at the Nile in 1798, and then how he ultimately helped bequeath it to those who followed after his death at Trafalgar.

* * *

The naval engagement off Cape Trafalgar is often presented as the culmination of an almost two-year-long naval campaign. In this sense it appears much in the same way as its land power counterpart, Austerlitz, an overwhelming defeat that decided not only the tactical, but strategic, issues that had been in play from the start. Nothing could be further from the truth. The great Chinese philosopher of war Sun Tzu (Sunzi) has said that "what is of supreme importance in war is to attack the enemy's strategy"[7] It is no exaggeration to say that the Royal Navy and British diplomacy—with not a little help from Napoleon himself—had undone Napoleon's maritime strategy almost two months *before* Trafalgar. Nelson's victory was not the reason that Napoleon cancelled his invasion of Great Britain in 1805—Napoleon's inability to bring his fleet together to escort his invasion flotilla across the English Channel was instead the result of a highly sophisticated and aggressive blockade by the Royal Navy.

THE PROTAGONISTS

The chief players in the naval campaign of 1803–1805 were the French and Spanish navies on the one hand and the Royal Navy on the other. It is no exaggeration to say that the Royal Navy had established a tradition of dominance over both these navies during the preceding 100 years—having lost only a handful of engagements, most minor.[8] During the period of the Wars of the French Revolution (1792–1802), the British had not only maintained this tradition but also improved upon it. In every major naval engagement during the period the Royal Navy not only won but also won decisively. The names of the great naval victories prior to Trafalgar roll off the tongue almost effortlessly:

- The Glorious First of June (Admiral Richard Howe, 1794, versus the French)[9]
- Cape St. Vincent (Admiral Sir John Jervis, 1797, versus the Spanish)
- Camperdown (Admiral Adam Duncan, 1797, versus the Dutch)
- The Nile (Admiral Horatio Nelson, 1798 versus the French)
- Copenhagen (Admirals Home Popham and Horatio Nelson, 1800, versus the Danish)

As can be seen, the constant here was British—not just Nelsonian—excellence and dominance over the other major naval powers (and not just the French). The British strategic problem was to ensure that its enemies' navies (composed of large ships-of-the-line or battleships) remained isolated and unable to mass superior force against the British.[10] The best way to do this was with a blockade (or a preemptive attack as in the case of the Danes). If breakout and unification were attempted, the British actively sought battle to restore the blockade. Herbert Rosinski, a famous professor at both the German and American Naval War Colleges in the 20th Century, wrote that

> In all narrow seas such as the Baltic, the North Sea, the Mediterranean or the West Indies, wherever fleets face each other at easy striking distance, there is and has been one and one way only for a naval commander to ensure the safety of his charges, and that is by sweeping his opponent from the board altogether obtaining against him undivided "command of the sea." Nothing short of that will suffice.[11]

Rosinski concluded: "Nor is blockade merely a way to acquire 'command'—by forcing the enemy to come out eventually and offer himself to battle—but it must be considered itself a *form* of command, because . . . it fulfills its function of excluding our opponent from the use of the sea [emphasis original]."[12]

J. C. Wylie, an American naval officer who was indirectly influenced by Rosinski, characterizes this sort of strategy as a *cumulative* strategy, and contended that attempts to employ what he called a *sequential* strategy against it are often inadequate, especially if the primary power is attempting to use land power against a sea power that does not share a land frontier with it.[13] Cumulative strategies often employ protracted, distributed military operations whose cumulative result denies an enemy something valuable while often conferring to its practitioner something valuable in return. In the British case, their dominance at sea allowed them to deny their enemies the use of the sea for all but the most limited naval actions while at the same time retain the use of the sea as maneuver space against continental opponents. When Britain committed her small army in Spain and Portugal, this was a great advantage indeed (see chapter 6). These insights of the naval theorists apply to Napoleon's attempt to counter Britain's dominance at sea by invading Britain with his army and achieving the decision ashore, not at sea. But first he had to cross less than 30 miles of ocean to do it. This he was never able to do.

The key to British dominance lay in the combination of several factors. The first of these is training, or more appropriately, seamanship and gunnery training. This factor was closely tied to the second, which has to do with the British tendency to aggressively blockade enemy fleets in port and actively seek battle with adversaries at sea. Because of its operational policy of being a sea-going and active fleet, the Royal Navy naturally had highly developed skills in both navigation and seamanship. Additionally, the Royal Navy, and in particular, leaders like John Jervis, Horatio Nelson, and Samuel Hood, realized the advantage they had in being able to practice extensively with their guns in the unfettered environment of the open sea—a skill much harder to practice in the calm, cluttered waters of a port. Gunnery in that age involved an almost ballet-like level of orchestration and the British had managed to accomplish a superior firing rate over every other navy, but only through continual and relentless training and practice with live ammunition.

These skills, superior seamanship and gunnery, only got better as those of the enemy declined with the maintenance of a blockade. When battle did occur, the British often had more maneuverability and the enemy fleet often had the difficult choice of either accepting battle or running away. Oftentimes, British seamanship, particularly if the British ships had the advantage of being upwind (colloquially known as "having the weather gage"), made running away a difficult option.[14] If battle was accepted, the opponents of the British now had to face, even if they had superior numbers, a struggle with an enemy that could outgun them and often fire three or more volleys of heavy cannon fire for every one of their own—a daunting prospect.

This disparity, almost asymmetry, in seamanship and gunnery led the Spanish and (especially) the French to develop countertactics over a period of generations. By the time of the French and Napoleonic wars, the French and Spanish were designing larger and often faster ships than their British counterparts. Speed enabled them to refuse battle more readily while the heavier weight of gunnery, it was hoped, would help balance the British rate-of-fire superiority. Further, the French, in particular, trained extensively on the accuracy of their long-range gunnery and focused on achieving what is referred to in today's military terminology as a *mobility kill.* The French would aim at the British ships' masts, rigging, and helm in order to disable their ability to close with them, or if closed, to get a maneuver advantage in order to negate British firepower in a slugging match.

In response, the British emphasized closing with the enemy fleet and boarding its ships if possible. Additionally, British gunners were trained to fire at the enemy hull in order to kill gun crews and personnel in preparation for boarding, and failing that, to sink the enemy vessel—but the goal was to try to capture the enemy vessel intact and then add it to the Royal Navy. The British also equipped their ships with the more destructive guns of the carronade type, which were meant for use at close range and used a composite load consisting of a large round shot and a keg filled with hundreds of musket balls. Nelson's flagship often carried two of the largest type of carronade, which fired a 68-pound solid round shot combined with a keg loaded with 500 musket balls—clear the decks for action indeed! British long-range gunnery was thus not as excellent on its ships of the line (battleships) but rather on its smaller screening ships such as frigates and brigs.[15]

Finally, the issue of leadership and morale must be addressed. Despite the draconian discipline in the Royal Navy, much of which would mutiny at Spithead in the spring of 1797, morale in the Royal Navy was often superior to that of its opponents. Clearly this was the case during all of Nelson's victories. This morale advantage was the result of a tradition of maritime victory and uniformly competent, and often superb, leadership. Another explanation can be found in the fact that the Royal Navy, despite its name, was not so much the King's Navy as it was Great Britain's navy. Just as the French Revolution had given the people of France a national army, so had the long evolution of the Royal Navy after the Glorious Revolution of 1688 given Great Britain a national or people's navy. The Royal Navy was a strong national institution by the time of the French Revolution.

Not only this, but as an institution the Royal Navy's officer corps consisted of a meritocracy, just as the French army had become after the Revolution. A career as an officer in the Royal Navy, while still related to patronage and privilege, was open to talent. The weak, stupid, lazy, and incompetent were, as often as not, weeded out by the physically harsh and difficult life. The officer corps for the Royal Navy came from a broader segment of the British population than did that of the British army—which was still very much the King's army even though it was no longer called the Royal Army (because it had revolted against the King during the English Civil War). One example will suffice—Nelson. Horace, later Horatio, Nelson was squarely from the middle class, being the son of an Anglican preacher (parson). His title was earned in battle through merit and not hereditary.[16] Napoleon employed precisely the same system once he became emperor, creating a nobility of merit.

NELSON'S APPRENTICESHIP IN OPERATIONAL ART—THE BATTLE OF CAPE ST. VINCENT

As we have seen, by 1797, the War of the First Coalition was going very badly for Austria, Great Britain, and their other minor allies. Prussia and Spain had left the war and in 1796 the Spanish decided to see if an alliance with the French might help them lay British pride low via the mechanism of sea power as well as lingering Spanish resentment of Britain's expansion of her oversea empire, often at Spain's expense, whether she was a neutral or an ally. Recall, too, that Spain had joined France during the American War of Independence, leading to her acquisition of Louisiana. More might again be gained since it appeared Britain was on the losing side. Spain's belligerence could not have come at a worse time for Britain. With France triumphant in Italy and threatening gains in Germany, the combination of the Spanish, French, and now Dutch fleets might finally offer Britain's allies the chance to combine and perhaps escort an invading armada across the English Channel.

Britain had held her own at sea, with the famous Admiral Richard "Black Dick" Howe punishing the French, in the battle on the "Glorious First of June." But French sea power remained, as Sir Julian Corbett termed it, "in being." If it could combine with new allies and the newly captured Dutch fleet, it might yet prevail. The Royal Navy had a huge job in 1797; it had to watch the French ports and fleets as well as those of Spain and the low countries where the Dutch fleet was gathered. It was actually a greater crisis than was to occur during the days of Trafalgar. However, British operations hinged on aggressive blockade and engagement. As Rosinski has written, the goal was to "sweep" the enemy fleet "from the board." However, the Royal Navy had to remain ever vigilant, on patrol, and constantly seeking an opportunity to catch the enemy unawares if they left port. The British mariners had reached a point that if offered an engagement, even if outnumbered, they would engage in battle. Such battles often had inconclusive results because all too often the French and the Spanish, in their superior ships, either returned to port or ran away—they could refuse battle in most situations unless caught by surprise. It was just this situation that offered the commander of the British Mediterranean Fleet, Admiral Sir John Jervis, the chance to catch the bulk of the Spanish fleet. Jervis found the Spaniards escorting a convoy of mercury needed in Spain's South American colonies to process silver at sea and precipitated an engagement of Cape St. Vincent in Portugal in February 1797.

Jervis was the ideal pick to command the fleet watching the Spanish. He had fought with General James Wolfe at Quebec, was aggressive, and a keen judge of character and talent in subordinates. Also, most importantly, when he recognized talent in his subordinates he was very enlightened about developing it and then delegating authority to the men he trusted—a key attribute of the philosophy of mission command.[17] Jervis's presence "transformed the spirit of the Mediterranean Fleet." When Captain Horatio Nelson, the second youngest British officer ever to make post captain, was assigned to his fleet, Jervis immediately took to the fiery, intelligent young officer. Jervis also promoted Nelson to the most senior rank possible for a captain, commodore, since Nelson now routinely commanded more ships than just his own. As a result, Nelson would command the rear division of Jervis's fleet, flying his flag aboard the battleship *HMS Captain* (74).[18]

As discussed, Jervis took over the Mediterranean Fleet at a critical juncture. With powerful Spanish and French fleets in play, the British pulled out of the Mediterranean as the Austrians suffered repeated blows from Bonaparte in Italy and pressure on the Rhine in early 1797. The war now assumed a principally naval character and Jervis's fleet, including Nelson, now concentrated solely on the Spanish threat to combine with the French and ferry victorious French legions across the channel. However, a truce with Austria that would release French troops was still two months distant (see chapter 2).[19]

One of the last groups of British ships to leave the Mediterranean belonged to Nelson. Temporarily in command of two captured French frigates (*La Minerve* and *Blanche*), he had miraculously sailed through the main Spanish fleet at night in the fog. He then proceeded directly to Admiral Jervis whom he found on February 13, 1797 off Cape St. Vincent on the Portuguese coast (see map 12). He gave Jervis the critical news that the Spanish fleet had entered the Atlantic. The next day, Jervis made contact with the Spanish ships under the command of Admiral Don Jose de Cordoba. Cordoba outnumbered Jervis by almost 2–1 in battleships (27 to 15). Jervis formed line of battle and made straight for the approaching Spanish fleet, saying "The die is cast and if there are 50 sail of line, I will go through them." This famous utterance highlights the British method of seeking battle in almost any situation except the most unfavorable. The Spanish fleet was still divided into two groups and trying to mass, one to windward (upwind) and one leeward (downwind) from the British. Nelson with his good friend Cuthbert Collingwood brought up the rear in *Captain* and *Excellent* (64).[20]

At the key moment in the battle, Nelson's portion of the line was closer to the Spanish than the head of the column as Jervis took his ships into a turn to keep the

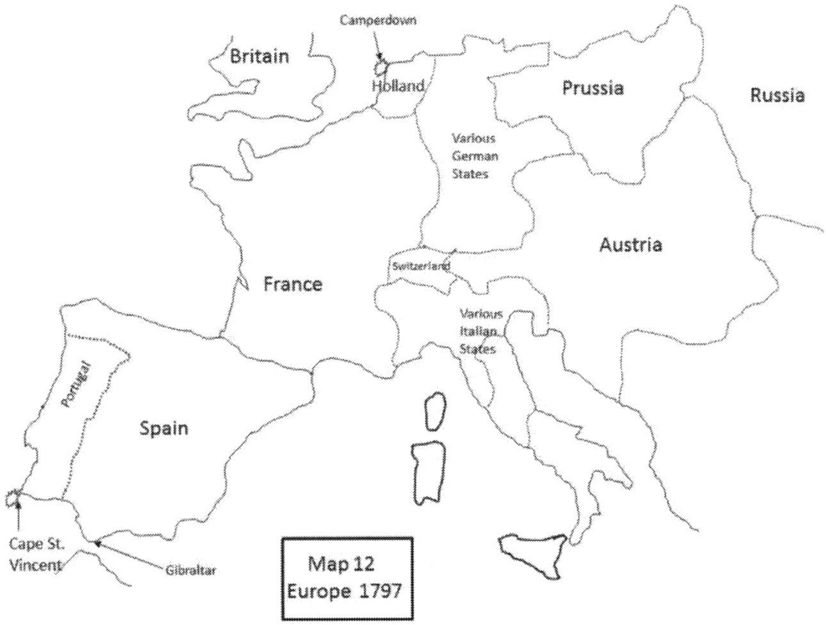

Spanish divided. It was at this moment that the British style of mission command evidenced itself. Nelson, disobeying the famous standing battle orders, wore out of the line and sailed straight for the middle of the Spanish column. This was where the most powerful Spanish ships were positioned, including Cordoba's massive flagship *Santissima Trinidad* (140). Nelson's aggressiveness and ability to act independently had been known in the fleet, but now they were on display for all to see. Jervis aboard the flagship *Victory* (100) saw Nelson engage seven enemy battleships with his one. He approved the action and signaled Collingwood aboard *Excellent* to support him. He then sent out the same signal to the remainder of his ships that Nelson himself sent from *Victory* at Trafalgar, "Engage the enemy more closely"[21] (see figure 1).

Jervis's signal was the essence of decentralized execution, leaving to each captain his choice on how and which Spanish ship to approach. Nelson's ship should have been obliterated as he endured the close but inaccurate fire of the Spanish. Elated by close combat, he rammed *San Nicholas* (80), which had become fouled close aboard with another Spanish battleship *San Josef* (112) and boarded her. Nelson personally led the boarding party aboard *San Nicholas* and took her in violent close quarters fighting. The next action he took was unprecedented and sealed his fame—with *San Josef* still close aboard, he continued with his boarding party across the captured ship

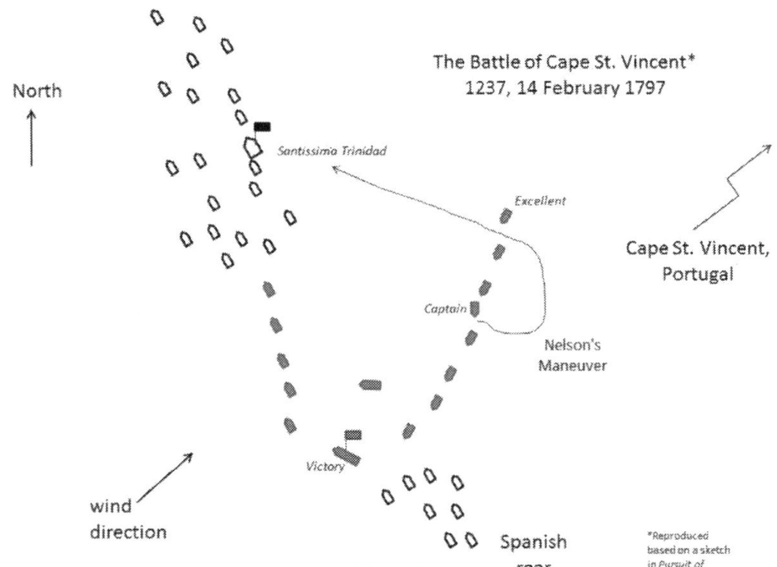

Figure 1

and boarded and took the larger ship from his new prize. Meanwhile, Collingwood had pounded three more Spanish ships to pieces and taken one of them. Of the four Spanish ships taken, two belonged to Nelson. Cordoba's flagship, the gigantic *Santissima Trinidad* had struck her colors, too, but was rescued by several other Spanish ships before the British could board her. Many of the Spanish ships that survived were badly damaged and thousands of valuable gun crews and sailors had been killed by the deadly British fires. The Battle of Cape St. Vincent highlights how absolutely decisive general engagements at sea can be, although they are much rarer than land battles because either belligerent often can refuse battle. In fact the Spanish were attempting to do just that when Nelson made his famous maneuver. It was a spectacular victory and earned Nelson a knighthood (of the Bath) and promotion to rear admiral. The King made Jervis Earl of St. Vincent in honor of the great victory, and Spanish sea power remained cowed until peace was signed in 1802 at Amiens.[22]

There still remained French and Dutch fleets to fight as well as the Nore and Spithead mutinies of 1797. So great was British ascendancy that the Dutch fleet remained in port during the worst moments of the mutiny when Admiral Adam Duncan kept watch with two ships, deceiving the Dutch by signaling a fleet over the horizon that was not there.[23] Once the mutinies were resolved, Admiral Duncan managed to demolish the Dutch fleet as a threat at Camperdown (October 11, 1797).

This was a victory nearly as great as Trafalgar with the British capturing 13 enemy vessels and emphasizes the general excellence of Britain's operational commanders at sea across the spectrum. Nelson, in the meantime, had been sent by Jervis on his first independent assignment as an admiral to seize the port of Santa Cruz on Tenerife in the Canary Islands. Disaster resulted. He made the classic mistake of underestimating his foes and overestimating his own power to prevail in an amphibious assault. He led his landing force into an ambush and Nelson almost bled to death when his right arm was shattered by a musket ball and later amputated.[24] The wound did not heal properly and Nelson returned to England as an opium-sedated wreck who believed his career had ended with the loss of his right arm, writing to a friend, "I am become a burthen to my friends and useless to my Country."[25]

NELSON IN OPERATIONAL COMMAND AT THE NILE

Although 1797 had seen Austria withdraw from the First Coalition, the British had calmed the worst fears of disaster by decisively defeating the fleets of France's two allies—the Spanish and Dutch. With two fleets "off the board," there still remained the most dangerous fleet of all—the French. Napoleon Bonaparte had proposed to the Directory that the best way to attack Great Britain was to threaten her lines of communication through the Levant with her Indian colonies, the crown jewel of the empire. To do this he planned to take an army composed primarily of veterans from his Italian campaign and sail to Egypt where he would land and establish a French colony, placing him astride the quickest British route to India. It was an ambitious plan, but in order for it to work the French had to slip by the Royal Navy. The Directory wanted to get rid of the political threat posed by Bonaparte and agreed to dispatch him with over 44,000 troops and the entire French Mediterranean Fleet to accomplish this task.[26]

One finds all the elements of mission command in Nelson's pursuit and annihilation of the French fleet that escorted Napoleon to Egypt, although his goal had been to sink Bonaparte's army while still embarked on its transports. Now in command of his own squadron, Nelson embodied for the Royal Navy the principles we have discussed but now with the new dimension of a command philosophy that empowered the tactical and operational initiative of his subordinates. His subordinates in turn relied on the teamwork and initiative of their crews. Similarly, we find that Nelson's commander, Admiral John Jervis, also exercising operational mission command in how he "controlled" his talented subordinate. The campaign of the Nile thus demonstrates how the Royal Navy's command philosophy at all echelons of command was a profound force multiplier. The period prior to the actual engagement in August 1798 has much to teach us about what Nelson himself called the "The Nelson Touch."[27]

Nelson recovered rapidly after his disaster at Tenerife under the close care of his wife Fanny, and, less one arm, he returned to command. Nelson raised his flag on *Vanguard* (74) in late March 1798 and set sail for the Gulf of Cadiz. Admiral Jervis's confidence in the younger admiral was unbounded. As soon as Nelson arrived to join the fleet on blockade duty off Cadiz, Jervis detached him on an independent command with a small squadron to enter the Mediterranean and keep an eye on the French fleet in Toulon. Jervis forwarded another 11 ships of the line to Nelson in May 1798, an unprecedented command for such a junior admiral, which caused much grumbling among the many admirals senior to him without sea command. Jervis instructed Nelson to stop Bonaparte's invasion force, suspected to be bound for Egypt. Nelson's plan was simple, to intercept the invasion fleet and destroy Napoleon and his army at sea.[28]

Meanwhile, Bonaparte's invasion armada had set sail from multiple Mediterranean ports that May, but unknown to Nelson it proceeded first to Malta, which it captured on June 10, and then proceeded leisurely to Egypt. Bonaparte was extremely lucky, as a fortuitous storm damaged Nelson's flagship and delayed his pursuit, while at the same time, the Malta diversion confused things a bit more. The slowness of the lumbering French armada probably saved it from being intercepted by Nelson at sea. Bonaparte landed at undefended Aboukir Bay on July 1–3 and then advanced and defeated the local Mameluks at the famous Battle of the Pyramids (July 21).[29]

Jervis's faith in Nelson was rewarded, but not immediately. Nelson's impatience almost did him in, but he was tenacious in pursuit of his quarry. As noted, on May 19, Bonaparte, escorted by the French fleet under Admiral Francois-Paul Brueys, had departed while Nelson collected his squadron after the storm. Nelson was desperately short of frigates (he had only three) to provide him intelligence and decided Napoleon's destination must be Naples. Bonaparte instead went to Malta and conquered it, as we have seen. Nelson realized his mistake and decided that Napoleon's next objective was Alexandria, Egypt. He rushed off arriving in the vicinity of Alexandria on June 29 and found nothing. Brueys and Napoleon had taken a different route north of Crete and sailed far slower than Nelson imagined. Nelson missed a great chance by not waiting off Alexandria, second-guessing himself and sailing north to Turkey. While Nelson sailed north, the French arrived and began to debark their troops. It seemed that Nelson had lost the game of cat and mouse and missed a golden opportunity to destroy both a French army *and* a French fleet.[30]

Nelson, like all great operational artists, did not give up. Off Sicily he learned of his mistake and doubled back as fast as his ships could sail to Alexandria. Late on August 1, he arrived and found the transports empty, but Brueys' 13 battleships and many smaller warships were still anchored close in to shore in the shallow and treacherous waters of Aboukir Bay. Brueys had unwisely sent half of his gun crews ashore to assist Napoleon with the land campaign. He also thought himself unassailable so close in to the shore. The final nail in his coffin was the late hour of the day. Surely Nelson would not attack in such dangerous waters with the onset of night? Nelson, demonstrating his intuitive grasp of the weakness of the French, instantly decided to attack.[31]

Nelson had planned in advance for this moment, and so made the decision almost instantaneously to attack the vulnerable French ships. To understand how he had operationally prepared one must go back many years. First, Nelson actively fought for the welfare of his sailors from the beginning of his career until his death in combat in 1805. The sailors who served under him knew this, and their support generated a confidence in his decision-making and leadership. Second, Nelson's rapid and seemingly "snap" decision-making in combat was not so much luck or intuition, although these certainly played a role, but more often "achieved by decisions made in the quiet of his cabin."[32] In other words, Nelson prepared himself intellectually at all times for decision-making, tactical and otherwise, through reflection during the often lengthy downtime that one experiences at sea (something the author of this book can attest to).

For this battle, Nelson had the complete confidence of his 15 battleship captains and this was rewarded by his confidence in them. Nelson also had the advantage of probably having some of the best captains, ships, and crews in the entire Royal Navy under his command at this battle. The 10 battleships that St. Vincent sent him under Sir Thomas Troubridge—who had served as a midshipman with Nelson—were later referred to by Nelson himself as, "the finest Squadron that ever sailed the Ocean." He also knew most of their captains personally, having served with most of them in combat, knowing only three by reputation alone. He soon remedied this deficiency by proactive efforts. After barely missing the French at Alexandria and as he sailed about looking for them, Nelson instituted the policy of bringing his captains individually aboard the flagship to mess with him and to share their ideas with him and he with them. In the words of a contemporary observer, "he would fully develop them to his own ideas of the different and best modes of Attack . . . they could [as a result] ascertain with precision what were the ideas and intentions of their Commander without the aid of further instructions." This, in short, epitomizes operational mission command, especially in a battle at sea where directing one's subordinate in the heat of an engagement often proved difficult if not impossible.

On June 22, Nelson brought his four most senior captains aboard *Vanguard* and asked them for their opinion on which way the French might have gone. They voted for Alexandria—they were all thinking along the same lines as Nelson but this shows he trusted in their judgment. On July 17, just before turning the squadron back to Alexandria, Nelson brought all 14 captains to the flagship to ensure commonality of purpose and a shared understanding for what he expected from them in the battle he was convinced would occur soon—if he could only find the French fleet. Nelson's only real "error" prior to the battle involved his reluctance to formally appoint a second in command since he was actually junior to several of them and did not want to upset the cohesion of the team in making one of them the object of jealousy by the others he did not pick. Perhaps he was thinking that if he were killed (a distinct possibility given his habit of being in the thick of battle), they would all continue to do their duty out of a sense of obligation to him rather than have their own fighting trim diminished by petty jealousy between themselves after he was gone.

At the time of the battle, Nelson had 14 battleships to 13 of the French, but the French had the heavier weight of gunnery (bigger ships and bigger guns). Upon sighting the enemy fleet Nelson issued signals 53 and 54 from the Royal Navy's official signal books. The first simply alerted all the crews of his fleet to "prepare for battle." The second directed them to be ready to anchor at a moment's notice by the stern of the ship, an evolution requiring the rigging of a very large anchor and cable at the rear of each ship since anchoring by the stern was not something the ships were normally rigged for. It meant, as well, that each captain and his crew would have to do this under enemy fire, during which time they would be unable to fire their guns while at the same time would have to furl (take down) the full load of sail while sailing in shallow waters. Obviously, a cohesive team and shared understanding had to be present for this unprecedented tactic to be executed properly. Its purpose was to suddenly stop the forward progress of the British ships so they could fire a full broadside into the French ships without waiting for the ship to stop moving, as one would when anchoring by the bow. Each captain, and more importantly their crews, knew exactly the risk inherent in this maneuver and welcomed it. Nelson's ships did not waste time signaling an acknowledgment or request back to Nelson for any further guidance. This was because they all knew their commander intimately, they knew their ships, and they knew that their enemy was unprepared—not only for battle but especially for this unprecedented maneuver. Nelson intended for this to take place in a single battle line between the French line and the open sea.[33]

The second, we might say graduate, level of this decentralized-type command emerged in what the captains then did on their own initiative. Nelson effectively gave his captains free rein when he sent the signal to his ships at 1730 hours (5:30 p.m.) "to form line as most convenient," leaving it to their discretion how to approach the French fleet since he was in the rear aboard *Vanguard*. The execution was not without some hiccups, such as when Captain Troubridge on *Culloden* ran aground just north of the French, but this also served to let the ships behind know where not to sail. Several ships performed the anchoring maneuver inexpertly and paid with heavy casualties as a result of being out of position. But most of Nelson's other captains served him well. The aggressive captain of the lead ship *Goliath* was Sir Thomas Foley. He instantly made the decision to go behind the first French ships and inside the French line. About half the other captains followed his initiative while Nelson signaled the remainder in the last bit of daylight to follow *Majestic* (74) on the northern side (see figure 2). This resulted in a double envelopment of the French ships in the first part of the line. Their mates further down the line were anchored solidly and could not help them.[34]

To make matters worse, some of the French ships had not strung cable between their neighbors to counter the known British tactic of "breaking the line" and then shooting longitudinally into the aft and bows of the anchored French ships. Again, taking local tactical initiative, several British ships performed this maneuver, especially *Leander* (50) and *Alexander* (74), dealing out further bloody devastation against the hapless French ships that were their targets. The night sky of Alexandria was lit up with the din and spectacle of burning French ships as Nelson pounded the French fleet to pieces while his ships proceeded down the line. Brueys' flagship *L'Orient* (120), surrounded on two sides in this manner and the second largest battleship in the world, blew up after absorbing incredible punishment, including combustible grenades that Captain Alexander Ball of the *Alexander* had specially prepared to set French ships afire if he could get close enough. Only Admiral Pierre Villeneuve in the rear escaped with two battleships and two frigates. The rest of the French fleet—including 11 battleships—was destroyed or taken. Nelson, in the thick of the fighting as usual, received a nasty head wound. [35]

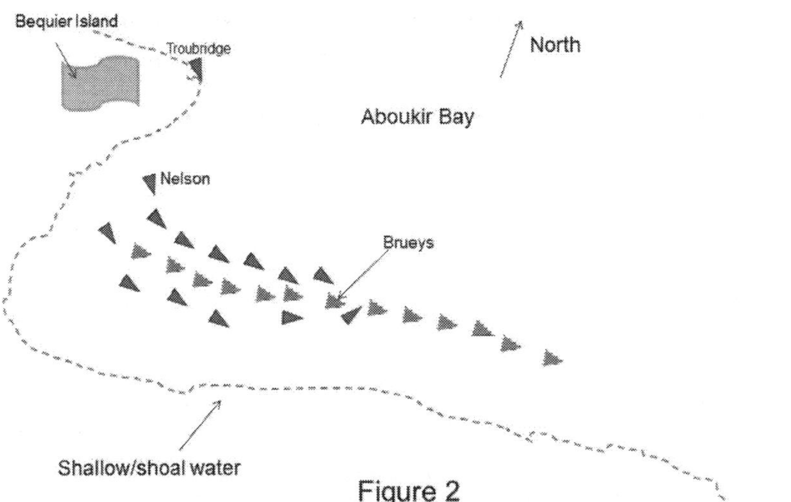

Figure 2
Simplified Sketch of Nelson's Battle of the Nile, August 1-2, 1798

Nelson had not lost a single ship and had suffered fewer than 900 casualties, most of these in ships that had mishandled the anchoring maneuver. He had effectively checkmated Napoleon's strategy by stranding his army in Egypt. When Napoleon attempted to fight his way through Ottoman (Turkish) territory in order to gain passage to Europe or even India, he was confounded again by another naval officer, Sir Sydney Smith, who successfully defended the Turkish fort at Acre, forcing Napoleon to return to Egypt. Later that year, Napoleon slipped through Nelson's blockade, abandoned his army, and returned to France where he established the Consulate with himself as first consul.[36]

THE NAVAL CAMPAIGN OF 1803 TO AUGUST OF 1805

The renewal of war between Britain and France, which most accounts begin with the year 1805, actually began in May of 1803 with a rupture over multiple unresolved issues. These stemmed in part from the flawed Peace of Amiens signed some 15 months before.[37] On May 14, 1803, Great Britain declared war on France. Nelson was appointed commander in chief of the Mediterranean Fleet assigned to watch the squadron of Admiral Latouche-Treville the same day. Two days later, he boarded *Victory* (100), Jervis's flagship at St. Vincent, and hoisted his flag.[38] He was back with the fleet he loved and that loved him.

Napoleon's strategy remained focused on gaining temporary command of the narrow seas of the English Channel in order to cross with an army of over 150,000 of the finest troops in Europe (*Le Grande Armée*). However, gaining command of the sea meant that the French navy must defeat or evade the Royal Navy—or perhaps deceive it and draw enough of it away—so that the invasion would have a reasonable chance of success. Simply put, the French strategy relied on an invasion that would enable Napoleon to dictate terms to a conquered British nation. The British problem was now exacerbated by the reemergence of a capable Spanish fleet, which upon Spain's declaration of war against Britain in December 1804 was added to that of the French. British strategy remained as it ever was, to blockade the French and Spanish navies in their ports so they could not escort an invasion flotilla.[39]

On August 26, 1805, Imperial French Headquarters issued orders to the *Grande Armée* (now some 210,000 troops) to turn east and abandon its positions along the English Channel for the invasion of Great Britain. Napoleon's decision to do this had occurred even earlier. Why did Napoleon abandon his attempt for a decisive conclusion to his two-year-old war with Great Britain by direct invasion? Since Trafalgar was some two months distant in the future one can categorically state that naval defeat was not the reason. The Royal Navy's blockade was the decisive element that defeated Napoleon's strategy—given its art and elegance it was surely a military operation that the subtle Sun Tzu would have appreciated. The British strategy accomplished its successful result not with a phantasmagorical naval battle for the ages, but rather with the slow and steady application of superior sea power over a period of two years. This in turn was augmented by the relentless offensive energy of admirals like Nelson, who literally hunted any French or Spanish ships that managed to slip by the blockade into the open sea. In the end, it was a combination of the systematic effects of the British blockade combined with the superior training, tactics, and leadership of the Royal Navy that accomplished this result.

One must give both the French navy and Napoleon due credit for operationally creating the conditions that might have given the invasion force a fighting chance. The great operational problem for the French and Spanish fleets was their inability to mass due to the blockade. If the majority of the separate fleets at Brest, Toulon, Cadiz, and other points could only combine, they might be able to defeat, or at least drive away, the British fleet guarding the English Channel. The only way to do this would be to try to split that fleet and then during the period of separation enter the Channel and do battle with the remainder. This would be difficult, since the British had large squadrons of battleships blockading all the major ports under a number of very competent admirals, Nelson, Cornwallis, and Keith in particular. The main British force—the Channel Fleet—was not even under Nelson but under Admiral William Cornwallis with another strong squadron under Admiral Keith in reserve. Nelson's task was to watch the French under Admiral Villeneuve, who were bottled up in the Mediterranean port of Toulon on the south coast of France. Villeneuve had replaced the more competent Latouche-Treville who had died in August 1804. Latouche-Treville's active defense strategy involved a war of attrition against Nelson's blockading squadrons by sailing out to sea and then dashing back to port, hoping to wear down his enemy while still maintaining the seamanship and morale of his crews. Unfortunately, the French admiral had a bad heart and died suddenly, although it is not certain that Napoleon would have left him in command given his caution and his preference for a protracted "fleet in being" defense whose results were not guaranteed. In any case, he kept the British on pins and needles for over a year and probably enabled Villeneuve's later success in slipping past Nelson's blockade to join the Spanish.[40]

Admiral Pierre Villeneuve had been in conflict with Nelson before—off the Nile. Recall that Villeneuve's *Guillaume Tell* was one of only two battleships to escape Nelson after the battle. Upon the resumption of hostilities in 1803, Villeneuve was in command of a squadron in the West Indies. He later commanded the French squadron based out of Rochefort on the French Atlantic coast. He was actually junior to Admirals Denis Decrés and Joseph Ganteaume. Ganteaume had the main French fleet at Brest. However, despite Villeneuve's junior status and deep respect and perhaps fear of Nelson, he was to be the primary instrument for obtaining command of the sea in the English Channel.[41]

Napoleon needed a way to diminish or remove the British naval dominance from the Channel and devised a daring plan to divide the British fleet. Nelson and the Admiralty believed that Napoleon would again try to disrupt their lines of communication to their key East Indies colonies by attacking either Malta or Egypt (as in 1798) as one way to distract the Royal Navy. Instead, Napoleon ordered Villeneuve to break out of Toulon and go west to threaten the British colonies in the West Indies, a tactic the French had employed with some success during the American Revolution. En route, Villeneuve would combine his fleet (11 battleships and 6,400 troops) with a Spanish fleet under Admiral Carlos Gravina. The French emperor hoped that Nelson would follow Villeneuve across the Atlantic. If all went well, Villeneuve would combine his force with whatever other French ships had broken out from Atlantic ports and made it to the Indies. In the meantime, the main fleet under Ganteaume would break out and proceed into the channel to cover the invasion. If Ganteaume did not escape on his own, Villeneuve's return with a combined Franco-Spanish fleet might be the agent for concentration of all three fleets, which would then press into the Channel for either battle or to cover the invasion, whichever occurred first. Unfortunately, Napoleon never sat his admirals down and explained the entire scheme to them, so that they often only knew the bare minimum of the details only after they had put to sea and opened their sealed orders. This was a recipe for disaster.[42]

Napoleon's plan, of which there were more than eight iterations, had many moving parts and relied on almost perfect synchronization between the various French and Spanish squadrons in order to mass in the correct place and at the correct time. It also relied on Nelson taking the bait and proceeding to the Indies in hot pursuit of Villeneuve. It came very close to success in achieving the desired concentration. By early 1805, many of the British ships had been at sea for over a year and were in need of repair and their crews were in need of rest. It was at this opportune time that Villeneuve made his first attempt to execute Napoleon's instructions. However, foul weather and the poor condition of the French ships and crews forced Villeneuve to return to Toulon. However, Villeneuve's first attempt had caused Nelson to make a grave error. He assumed that Villeneuve was sailing for Alexandria in Egypt as Napoleon had done in 1798. By February 7, 1805, Nelson and his 13 ships of the line had arrived off Alexandria, but they were in very poor condition. This was not only a result of their lengthy time at sea since the outbreak of war but also due to the Admiralty's failure to perform hull maintenance on many of them during the respite offered by the Treaty of Amiens.[43]

When the French fleet was not found, Nelson pressed back across the Mediterranean, still convinced that Villeneuve's objective was Egypt. It was because of this notion that Villeneuve's second attempt to break out and join with the Spanish succeeded. Nelson had precious few frigates to watch for Villeneuve and he chose to use these to watch the routes to the east and only one route to the west. Villeneuve cagily slipped out of Toulon on March 30 and then passed north of the Balearic Islands. By April 9, his slow-moving ships were in Cadiz and had combined with the Spanish. This combined fleet of 18 battleships and over 5,000 troops now proceeded to Martinique in the French West Indies, arriving a month later. Nelson had again proceeded east to pursue a phantom French fleet he thought was bound for Egypt. He was in the Central Mediterranean when he finally received word that Villeneuve was bound for the West Indies. Nelson immediately gave chase and by May 4, 1805, he was at the Strait of Gibraltar taking on stores for an Atlantic crossing. Villeneuve had a month head start and it seemed almost certain that Napoleon's plan must work. It was now that British seamanship and French indecision played their fateful roles. Villeneuve had secondary orders to attack Barbados but waited until June 9 to double back across the Atlantic and join up, hopefully, with Admiral Ganteaume to cross the Channel.[44]

Nelson and his fleet were in the meantime accomplishing the impossible, pressing across the Atlantic in three weeks in their leaky undermaintained ships—an act of incredible seamanship. They shaved off critical weeks from Villeneuve's head start. In fact, Nelson arrived on June 4 and was already hunting for Villeneuve. Once Villeneuve learned of Nelson's presence, he set course for the Bay of Biscay and a hoped-for juncture with Ganteaume. Although this was probably the right strategic move, it lowered the morale of the Franco-Spanish fleet, which only saw that it was again running away. Nelson soon learned of Villeneuve's departure and immediately dispatched the brig *Curieux* across the Atlantic to inform the Admiralty of Villeneuve's impending arrival, perhaps in the Channel itself. Luckily, the *Curieux* ran into the combined fleet and was able to report to the British Admiralty that Villeneuve was indeed heading for the Bay of Biscay. The British high command reacted decisively in positioning over 600 ships of various sizes to meet the Franco-Spanish threat. Again, superior British seamanship had prevailed. *Curieux* also passed word to the Admiralty of Nelson's own return (He had provisioned and set sail about a week after he dispatched the *Curieux*.) It was at this point that Nelson temporarily disappeared from the story.[45]

With Nelson's intelligence in hand, Admiral Cornwallis raised his blockade of Brest and dispatched a squadron of ships under Admiral Sir Robert Calder to intercept Villeneuve. Villeneuve collided with Calder in the fog off Cape Finisterre (northwestern Spain) on July 22 in a tactically indecisive action. Its results, however, were strategically decisive. Superior British gunnery caused the surrender of two French ships. This was enough to cause Villeneuve to pull back into the Spanish port of Vigo. Ganteaume, unaware that Cornwallis had raised the blockade, serenely remained in port so that by late July any hope of a juncture with Villeneuve was gone. Ironically, Calder was criticized for his strategic victory because he had not captured or sunk more enemy ships. He bore the stigma of having *not destroyed* Villeneuve's fleet for the rest of his career.[46]

Not long after these events, Nelson, again sailing his fleet across the Atlantic more quickly than a French frigate might, returned to the European waters. He had been constantly at sea for over two years, sailing over 10,000 nautical miles in 1805 alone. Villeneuve and Gravina remained united but blockaded in Cadiz (to which they had moved on August 20) and Ganteaume remained under blockade in Brest. Nelson now took the opportunity to try to repair his worn ships and rest his tired seamen while maintaining vigil off Cadiz, keeping his fleet just over the horizon (as was his habit) and maintaining watch with a string of frigates.[47]

Finally, Napoleon's aggressive actions in Europe had aroused jealousy and resentment among the other great powers of Europe. Napoleon's aggrandizing actions in no small measure contributed to the achievement of a long-term goal of British diplomacy—the formation a Third Coalition composed of Austria, Russia, Great Britain and several lesser powers to renew the continental conflict with France [see chapter 5]. Had this coalition not formed, Napoleon might have remained with this army along the English Channel. However, he now had threats to his strategic backdoor, as it were, and completely recast his strategy in the light of new realities. The long blockade operations that stymied Napoleon's strategy vis-à-vis Britain had contributed substantially to the favorable political environment of 1805. Nevertheless, the Royal Navy did not rest on its laurels, and Nelson, especially, planned for the destruction of Villeneuve's Franco-Spanish fleet.[48]

As he did before at the Nile, Nelson prepared carefully for battle. When he rejoined his fleet blockading the Franco-Spanish force off Cadiz in late September 1805 he immediately convened a meeting with his subordinate commanding officers onboard *Victory*. He had already sent a letter to all of them explaining his intention to attempt to trick Villeneuve into battle. Once battle was joined, he would attack with two columns of ships that would break the traditional battle line of the enemy into three groups: the van, the main body, and the rear. Of these groups, those in the van would be effectively unavailable for the initial phases of the fight since they would have to sail back into the wind (tack) to reach the battle, by which time Nelson intended to have defeated the other two groups of ships.

In this way Nelson would rectify any numerical superiority that the enemy might have as well as deny him the ability to refuse battle with the entire fleet (see figure 3). All of this was explained in further detail by Nelson in person aboard his flagship to his captains, usually at dinner. It was a supreme example of a commander making known his intent to his subordinates in a collaborative and accessible fashion. It bears many parallels to a similar dinner that Napoleon held with his marshals and officers on the night before Austerlitz.[49]

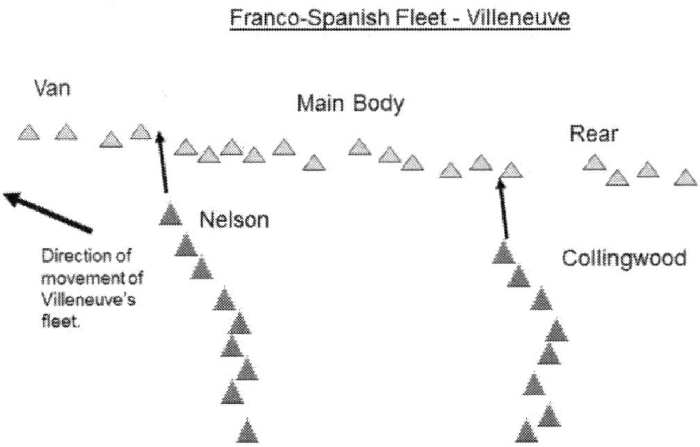

Figure 3
Simplified Sketch of Nelson's Battle Plan for Trafalgar, October 21, 1805

Nelson reiterated his intent in a famous memorandum written after his conferences and dinners aboard the *Victory* on October 9. Nelson's memorandum emphasized several keys to the victory that have already been mentioned: superior gunnery and seamanship. Nelson outlined for his captains that he did not plan to form new lines of battle once Villeneuve's fleet was sighted. In order to take advantage of superior British seamanship and trap Villeneuve he intended to waste no time forming up for battle but rather his ships' positions—what was called their order of sail—would also be their order of battle. This would give him speed. Second, he wrote that "no captain can do very wrong if he places his Ship alongside that of an Enemy." This instruction emphasized his belief that his fleet was ship for ship and man for man superior to that of the French and Spanish. It was also written to emphasize to his captains that if signals could not be seen from the flagship then they had complete autonomy to engage the enemy as they saw fit. Nothing Nelson ever wrote has emphasized the principle of decentralized execution as cogently. Finally, his intention was no mere victory but rather "annihilation" of the entire enemy fleet.[50]

Nelson now turned his mind to the battle readiness of his fleet and detached several battleships to escort a convoy of supplies for the fleet bound from Malta through the strait of Gibraltar. In fact, procuring fresh meat and vegetables for his scurvy-weakened sailors consumed much of Nelson's time as he attempted to get these items out to the fleet on station off Cadiz. Nelson's care with logistics is a hallmark of the operational commander who knows that healthy crews are fighting crews. Nelson also employed what we might today call operational deception.[51] Finding the fleet in a close blockade off Cadiz he decided to move it over the horizon so that the French and Spanish would not be aware of its precise location. He maintained watch on Cadiz with his frigates, which could also intercept and turn back snooping Spanish or French ships attempting to gather intelligence. Nelson meant to lure Villeneuve to sea, knowing that if the Frenchman could see his fleet he might not attempt a breakout for the Mediterranean. Once Villeneuve emerged, Nelson's intention was to intercept and destroy him, but until he judged the wind and time ripe he would remain just out of sight over the horizon paralleling Villeneuve's course.

Events conspired to cause Villeneuve to finally nerve himself to sortie from port. He knew the British force, through intelligence sources of his own, had been depleted by several battleships and was now inferior to his own. He also knew that if he did not sail, he might not have a fleet to command in any case since he knew a replacement was on the way to relieve him. Napoleon had removed Villeneuve from command but his successor (and rival) Admiral Rosily was still traveling via the overland route through Spain to Cadiz to take command.

Now all was in the hands of the weather, specifically the winds. On October 18, Villeneuve felt confident that the wind was favorable and ordered his fleet to unmoor. On October 19, the first ships began to leave the harbor, although the wind occasionally died, which caused Villeneuve to have several of his battleships towed out of port. Nelson's frigates signaled the good news that Villeneuve was finally leaving port, bound for Toulon and the Mediterranean in accordance with Napoleon's orders.[52]

Historians have argued that Trafalgar had been won before the first shot. This is probably hyperbole but contains much truth, nonetheless some hard fighting had to be done and so we return to Nelson's masterpiece to see how it all played out in execution.[53] Nelson had 27 battleships to 33 for Villeneuve, but this quantitative disadvantage was more than compensated for by the superior morale, leadership, planning, seamanship, and gunnery of Nelson's fleet. Nelson continued to stalk Villeneuve the next day as the Frenchman proceeded east-southeasterly toward Gibraltar with the combined fleet, however the pace of the French and Spanish advance was so slow that Nelson outran Villeneuve and had to bring his entire battle line around 360 degrees in order to fall back into a correct position vis-à-vis the combined fleet. Villeneuve by now had learned he was being stalked. Despite his numerical superiority at 7:30 A.M. of October 21, he ordered his fleet to "wear" back toward Cadiz and began a gradual turn of the entire fleet back to port.[54] He was literally running away from battle. This did not affect Nelson's plan appreciably; the only difference would be that Nelson's column would strike near the main body of the combined fleet's line while Collingwood's would now strike the rear.

Nelson had prepared for the light and variable winds that made the collision of the two fleets an agonizingly slow process. He had directed his captains to approach with all sail, including studding sails (auxiliary sails that take advantage of every last bit of available wind for motive force). In this way, Nelson would maximize his speed of approach to the enemy, which would minimize the amount of time his ships were under fire before they could respond. Once they pierced the enemy line of battle, the British sailors were to cut these sails away, which would have the double effect of slowing them down (for close engagement) and eliminating these sails from tangling up the standard rigging. Nelson also employed tactical deception by having his own column feint toward the enemy van of ships to confuse them as to his real intention and then turned at the last possible moment toward Villeneuve's flagship (*Bucentaure*) in the main body. The result of this action forced the van to maintain its line ahead and to delay its turning back to come to the aid of the rest of the fleet (see figure 3).[55]

Nelson then added a final edge to his fleet's fighting mettle when he sent, sequentially, two significant signals, neither of which were tactical but rather inspirational as a means to give his men a combat edge. The first signal elicited a spontaneous outbreak of cheering throughout the fleet: "England expects every man will do his duty." This signal was passed a quarter of an hour before the first gunfire and nearly five hours after Nelson had ordered the turn toward the enemy. He followed this with a signal for his captains: "Engage the enemy more closely." This signal might be regarded as Nelson's final turnover of the battle to the discretion of his captains. After this point, the British fleet was now in the execution phase of the battle, which relied relatively little on Nelson's personal direction and almost wholly on the actions of his subordinates.[56]

The last minutes of the approach were harrowing as the lead ships of each column traveled under the heavy and concentrated fire of the combined fleet. Nelson and *Victory* led the northwestern column and Collingwood was in *Royal Sovereign* (100) at the head of the southeastern one (the French were heading north toward Cadiz). This provides more evidence of the superiority of the British system since the loss of both ships would mean the loss of the two flag officers—however, they were no longer absolutely necessary for success in the overall battle and became more or less fighting members of the crew of the ship they were on. Collingwood won the race and *Royal Sovereign* crashed into the rear third of the French-Spanish fleet just past noon. He had minimized his casualties by having his sailors lie down during the final minutes of the approach so he would have every man possible to work the guns and deliver the first devastating broadsides, simultaneously, into the ships on his left and right as he pierced Villeneuve's line, as his captains had done at the Nile. The French and Spanish could only reply with the relatively few guns mounted on the bow and stern of their vessels.[57]

The battle now became a melee as Nelson's *Victory* plowed between Villeneuve's flagship *Bucentaure* (80) and the battleship behind her, *Redoubtable* (74). The French gunners had had much better success against *Victory*, which lost many of her sails and had had her helm shot away, yet her momentum carried her ahead as planned. *Victory*'s point-blank broadsides, especially from her 68-pound carronades, caused horrendous casualties in the two French ships. Not long after, a sharpshooter aboard *Redoubtable* (74) sighted the bemedaled Nelson, who had refused to take cover below decks, took a bead, and fired. The ball hit Nelson in the left shoulder and then ricocheted through his lung, finally lodging in his spine. This mortal combination of wounds led to Nelson's slow and painful death. *Redoubtable* now prepared to board the ailing *Victory* but her boarding party was wiped out by a devastating broadside from the *Téméraire* (98), following in line behind *Victory*.[58]

Much of the remainder of the battle followed this same pattern of devastating close combat, but in all cases the British had the better of it up and down the line. Meanwhile, Villeneuve was frantically signaling his van under the French Admiral Dumanoir to come about to assist in the battle. Dumanoir did not see the initial signals due to the smoke of battle and continued sailing toward Cadiz. It was only at 3:00 P.M. that he came about. By this time, about 15 ships of the combined fleet had struck their colors (surrendered), including Villeneuve aboard *Bucentaure*. Virtually all of them were blazing charnel houses full of dead and dying men. What remained of the French van, unable to affect the outcome of the battle, fled to Cadiz. Not a single British vessel had been lost. Nelson died around 4:30 P.M., but he died with the knowledge that his men had won the most complete victory at sea ever obtained by a modern sailing fleet.[59]

In total, the British captured 10 French and 10 Spanish ships, however, a subsequent terrific storm damaged or sank many of these and the British were only able to salvage 4 battleships to add to the Royal Navy's order of battle. British casualties at Trafalgar numbered 449 killed and 1,241 wounded. French and Spanish casualties, including prisoners, were almost 10 times that of the British, including over 4,000 killed alone. Several days later, British admiral Sir Richard Strachan intercepted, attacked, and captured four of the surviving French ships that had escaped the battle intact.[60] Thus the Royal Navy gained eight new battleships as a result of Nelson's and Strachan's battles.

THE LEGACY OF NELSON FOR NAVAL WARFARE

Prior to Nelson's famous victory, other British admirals had indeed mirrored and even suggested his approach of centralized planning and decentralized execution. One recalls especially Admiral Rodney whose melee tactics against the comte de

Grasse during the American Revolution foreshadowed those of Nelson.[61] However, the Royal Navy's institutional attitude toward Nelson's methods, best highlighted at the Nile, was only occasionally and arbitrarily applied by the other admirals of the fleet prior to that battle. After the Nile, these methods became the norm, not the least because it was Nelson's way as much as for their excellence from an operational standpoint. The results of subsequent history contradict those who argue for the uniqueness of the results of the "Nelson Touch" and its confinement only to the period of his executive leadership.

* * *

Despite Napoleon's best efforts to rebuild and reanimate his navy, the French never again came close to challenging Britain at sea. Nelson's heirs had no real difficulties maintaining their ascendance and even repeated Nelson's destruction of Danish sea power at Copenhagen in 1807, this time in concert with the army. By 1813, Napoleon's need for men and material, particularly trained artillerists and infantrymen, was so great that ships were denuded of their crews. These men were then formed into new batteries and *regiments du marine* that were promptly marched off to Central Germany to perish on land. The other great result of Trafalgar involved Napoleon's imposition of the Continental System as a cumulative economic strategy against the cumulative sea power strategy of Britain. From this "paper blockade," never realistically enforceable, flowed much that would result in Napoleon's downfall, including his disastrous interventions in the Iberian Peninsula (see chapter 6) and Russia in 1812 (see chapter 7).[62]

Trafalgar was decisive, but mostly for the period after 1805 and not so much for the events that occurred on land and sea in 1805. The British naval campaign of 1805 and Nelson's subsequent engagement at Trafalgar embodied the successful application of superior sea power by a major European power and highlight the *necessary* role that command of the sea by Great Britain would play in shaping the eventual defeat of Napoleon. Finally, Nelson's style of command demonstrated the superiority of the principles of centralized planning, decentralized execution, and operational mission command for the successful prosecution of modern naval warfare.

FIVE

NAPOLEON ASCENDANT

JOMINI AND CLAUSEWITZ IN HARMONY

CHRONOLOGY

Dec 1804	Napoleon crowned himself emperor of France. Spain declared war on Great Britain.
Aug 1805	Austria, Russia, and Great Britain form the Third Coalition against France.
Aug 26	The *Grande Armée* began to depart its cantonments along the English Channel and marched to the east.
Sep 24	The *Grande Armée* arrived along the Rhine.
Oct 21	Mack surrendered Austrian army at Ulm to Napoleon Nelson defeated Franco-Spanish fleet at Trafalgar.
Dec 2	Napoleon defeated Austrian and Russian forces at Austerlitz.
Oct 8, 1806	Prussia declared war on France; War of the Fourth Coalition began.
Oct 14, 1806	French armies defeated Prussian armies at Jena and Auerstadt.
Oct 25	French troops under Marshal Davout entered Berlin.
Nov 28	Napoleon entered Warsaw.
Feb 7–8, 1807	Napoleon fought to a standstill at Eylau by Russians and Prussians.
Jun 14	Napoleon defeated Russian army at Friedland.
July 7	Treaty of Tilsit between French and Russian Empires brought general peace to continental Europe.

> *We owe it to Napoleon . . . that we now confront a succession of a few warlike centuries that have no parallel in history . . . we have entered the classical age of war, of scientific and at the same time popular war on the largest scale (in weapons, talents, and discipline). All coming centuries will look back on it with envy and awe for its perfection.*
>
> Frederich Nietzsche[1]

This chapter examines the period of Napoleon's most dazzling campaigns with an instrument of war that bore little resemblance to anything that preceded it: the *Grande Armée*. The three continental campaign wars from 1805 to 1807 highlight the ascendancy of the French on land in the face of established British command of the seas. This period also illustrates the asymmetry of French arms against all other opponents as reflected in the brilliant campaigns of Ulm-Austerlitz, Prussia, in 1806, and the less brilliant, but victorious, campaigns in Poland in 1807 against the Prussian and Russian armies. The operational aspects of the writings of Jomini and Clausewitz owe much to this period of Napoleon's military ascendancy. Jomini is the apostle of the offensive and these campaigns were primary evidence for his position. At the same time, Clausewitz advances the notion of a form of war that approaches the "absolute" in actual execution and these campaigns also support this idea. We also find a third party, Frederich Nietzsche from a less optimistic age, commenting, as did both Clausewitz and Jomini in their way, on the awesome, terrible "perfection" of war as was practiced by Napoleon in these campaigns. All three men saw in these events the *operational* shape of things to come.

* * *

From 1803 to 1805, France and Britain fought a maritime war. By late 1804, however, the winds of war on the continent again began to blow. Tolstoy's opening line from *War and Peace* captures the Russian attitude as well as any: "Well, Prince, Genoa and Lucca are now become no more than private estates of the Bonaparte Family. No, I warn you, if you do not tell me we are at war . . . you are no longer my friend."[2] Tolstoy emphasizes how Napoleon continued the French policy of using peace as an extension of war to aggrandize the state—now officially recognized, by the French at any rate, as an empire. Bonaparte, tossing away pretense, had made himself Napoleon I of that empire. In addition to closing all ports under French control to British commerce, and those of Spain in late 1804, Napoleon's actions, especially his extra-judicial murder of the Duc D'Enghien after kidnapping him from a neutral area of Germany, led the great powers of Europe to form yet another coalition against him.[3] The young Tsar Alexander I made common cause, as had his father, with the Austrians, and accepted British financing to rid Europe of the French menace. Britain, after two years alone against the might of France, now had great continental armies available to pull Napoleon's *Grande*

Armée eastward from its cantonments around Boulogne and the other channel invasion ports.[4] As discussed in chapter 4, Napoleon's decision to abandon the invasion of England in 1805 occurred for this reason as well as the undoing of his complicated plan to break the British blockade—*not* the Battle of Trafalgar later that October.

It is to the *Grande Armée* that we must now turn. Napoleon had considerably refined the French military into an unprecedented operational tool—a field army of unprecedented size organized, commanded, and trained meticulously such that it could operate along multiple lines of operation in a distributed fashion on a very broad front. In short, the *maestro* operational artist Napoleon fashioned a tool with which to improvise his masterpieces, the brilliant and decisive campaigns of 1805–1809. The *Grande Armée* was this tool. First, we must understand that Napoleon inherited—or perhaps more accurately seized—a military system already molded to a rather high degree of competence and effectiveness in the crucible of the wars of the French Revolution. He had hundreds of thousands of mature, well-equipped veterans, led by experienced and competent officers, and a state that had been dedicated for 10 straight years to sustaining, training, and employing these military forces. Napoleon institutionalized these changes to the French military machine and then added additional combat power, principally through refinements in organization, education, and motivational psychology. It is to these refinements that we now turn our attention.

As outlined in chapter 2, the Napoleonic Wars constituted a combined arms revolution in tactics and doctrine. Napoleon continued this revolution, especially at the higher operational levels. Although previous generals, including Napoleon, had employed ad hoc corps, no one had formalized and institutionalized the use of multidivision formations. As the size of Napoleonic armies increased, the need to group divisions together to enhance both their survivability and combat power became obvious. However, a new level of tactical and operational command was needed for these previously ad hoc formations. Napoleon's genius in this regard was to simply keep doing what the French had done with the division, except at a higher echelon. As an army commander in Italy and elsewhere he had found that his operational span of control could only cover, effectively, no more than about seven or eight subunits. In order to solve this problem with field armies that now numbered a dozen or more divisions each, he needed an intermediate echelon between the division and the army. One might reasonably assume that the solution was simply to create many armies and parcel out these commands to half dozen or more commanding generals, which is what the Directory had done. However, Napoleon wanted to retain absolute unity of command over one gigantic army, into which he would put most of the Empire's combat power for his command and use only.[5] This meant he had to have an intermediate level between the division and the army to effectively control this gigantic mass. Napoleon did this by creating the *Corps d'Armée* or army corps as a standing military formation.

The Napoleonic *Corps d'Armée* varied in size, often in relation to the talents of the generals Napoleon chose to command them. His first innovation in this regard was the command element for these corps. During the period of the peace after Amiens he resurrected the French rank of marshal and awarded it to the most deserving and competent (and in some cases difficult) of his contemporary revolutionary generals. This did not mean he might not give command of a corps to a nonmarshal, but his marshals were almost always his first choice to command these units. Returning to the composition of these organizations, corps usually included anywhere from two to five divisions. They also had their own corps artillery separate from the artillery organic to the divisions, usually several batteries strong, including both howitzers and regular field guns. This allowed a corps commander to weight his defense or attack using these guns. The corps also included permanent assignment of light cavalry—initially the idea was to assign one light cavalry division to each corps, but later corps might only include one brigade. This cavalry's role was obvious, to screen the corps as well as provide operational security for it so that it might never be taken unawares. For the composition of the *Grande Armée* in 1805 Napoleon created seven numbered corps (I–VII) under his personal command.[6] The only other substantial field army was the old Army of Italy under his most reliable general for independent field command, Marshal Masséna, a force that included many Italian troops. For all intents and purposes, Masséna's army was really just a much larger army corps—although he would form ad hoc corps of his own as the situation dictated.[7]

Napoleon's other organizational innovations vis-à-vis the *Grande Armée* involved doing at the level of the army the same sort of thing he did with the corps. He created specialized operational reserves encompassing all three branches. The first reserve was an infantry heavy elite corps; his Imperial Guard, composed of his own *guides* from the Egyptian and later Marengo campaigns; and other Republican guard formations such as the Consular Guard. The Imperial Guard included the famous grenadiers and chasseurs in their fearsome bearskin hats as well as cavalry, initially mounted chasseurs and *grenadiers a cheval*. The Guard also included the renowned "beautiful daughters," 12-pounder field guns of the Guard Artillery, as well as special guard horse artillery. It was an elite combined arms corps that eventually grew into an army within an army. Napoleon rarely used it en masse, but was prepared to do so in extreme circumstances such as at Krasnoe (Krasny) in 1812 and Waterloo in 1815. The Guard was above all both a tactical and operational reserve and would serve in this latter function as a replacement army when Napoleon lost so much of his regular army in Germany in 1813.[8]

The second element that Napoleon created as a standing reserve was his artillery. This mostly consisted of the Guard Artillery mentioned earlier, but he often augmented it with artillery from other sources to achieve massive firepower at a decisive point. Its operational function was to give Napoleon tactical options in choosing engagements. It should not surprise us that Napoleon used artillery in this way and that the key to his use of it was concentration and timing. His enemies often outnumbered him in terms of numbers of guns, especially after the disastrous Russian campaign, but Napoleon's use of his artillery on a battlefield remained a wonder to all who observed it right to the end at Waterloo.[9]

Another operational component outside the Guard that Napoleon maintained was his Cavalry Reserve. The impetuous Joachim Murat, perhaps the most dashing (and flamboyantly dressed) cavalryman that ever lived, commanded this huge mounted force for most of its existence. By the time of the start of the 1805 campaign along the Danube, the French cavalry had become much improved. Napoleon converted the old Republican line cavalry regiments into helmeted cuirassier regiments, amalgamated together with two bear-skin-capped elite *Carbinier* regiments into several divisions with supporting horse artillery. He also amalgamated the dragoons and light cavalry (hussars and chasseurs) into separate dragoon and light cavalry divisions. Of particular interest in 1805 was his experiment of including with this reserve dismounted dragoons, whom officers like Generals Marie-Victor Latour-Maubourg and Louis Baraguey D'Hilliers attempted to resurrect as mounted infantry. The great problem here was that since there were so many dragoon regiments and only so many suitable mounts, a much larger and hardy horse being required to carry the considerable kit of the dragoons. As a result, almost half of these regiments served on foot as light infantry in 1805, initially to be remounted after the invasion of England. By 1806, Napoleon was finally able to mount them after capturing the excellent stocks of horses in Saxony and Prussia. By that point the dragoons had become a sort of medium-heavy cavalry, often being used in the same manner as the cuirassiers. For the 1805 campaign, this large force would number over 20,000 men and gave Napoleon many operational options, including using it independently if needed.[10] Finally, Napoleon could, and did, use entire *Corps d'Armée* as reserves, for use as opportunity dictated, especially if they had not been decisively engaged but were near enough for an operational pursuit.

At the top of this fearsome instrument of war, Napoleon placed himself, achieving absolute unity of command in strategy, military affairs, and the formulation of what we today would call national security policy. Napoleon informed this unity with a polymath-savant's brain—he ran this enterprise using the subtle genius of his computer-like mind and his almost inexhaustible capacity for work. His predecessors in this method of civil-military rule are few and constituted his role models—Frederick the Great; John Churchill, Duke of Marlborough; Gustavus Adolphus of Sweden; and Julius Caesar. Napoleon surrounded himself with a staff whose one purpose was the rapid transcription and delivery of his orders and collection of reports and intelligence from his subordinates and augmented with a group of highly trusted aides de camp who could and often did act on his authority at critical points on the battlefield. These men included A.J.M. René Savary, who acted during the 1805 campaign as both an aide and intelligence officer; the scrappy Jean Rapp; and the redoubtable Georges Mouton (later Count of Lobau), whom Napoleon referred to as his "lion." It was Europe's first modern general staff, but it would soon be eclipsed in the aftermath of Prussia's defeat in 1806 and it absolutely depended on the genius of Napoleon to function smoothly.[11]

Two other innovations must be discussed. The first was Napoleon's complete revamping and overhaul of French officer education. Much of great value in the old Royal education system had been swept away by the Revolution but Napoleon made wonderful progress in reestablishing a system of schools to train new officers as well as further educate field-grade leaders. However, all observers agree his incredible wastage of officers, especially junior ones, in his many bloody campaigns threatened to overwhelm the output of this educational system, a system that did not serve him as well as it might have had he had a period of peace to consolidate his reforms. The second innovation regarded his institutionalization, across all of society, of his system of merit in the creation of the legion of honor. Napoleon led the way, always wearing his own legion of honor ribbon prominently, as did any soldier or civil servant who was awarded one. Single-handedly he instituted a device that increased the combat performance of his troops using an institution of honor instead of relying on the older Republican call to virtue. He democratized reward, and it paid him handsomely again and again. Today's militaries owe their own systems of awards and commendation primarily to Napoleon.[12]

The final innovation for discussion involved Napoleon's revolution in military strategy, best highlighted by the campaigns of 1805 along the Danube, 1806 in Saxony and Prussia, and, to a lesser extent, in Poland in 1807. As one historian has noted, the *Grande Armée* gave Napoleon the power to "scramble" his way to victory because his enemies were unable to avoid being fixed and defeated by this new and powerful system of operations.[13] Like Suvorov, but on a scale never practiced by the Russian, Napoleon intended the largest possible operational meeting engagement—which sometimes became an entire campaign. This was Napoleon's idea of strategy, as conceived by both Clausewitz and Jomini, of choosing to devote the bulk of his military power to one theater of operations and then aiming relentlessly at the complete destruction of his adversaries' military forces in one campaign season. Clausewitz labeled this form of war *the higher form* or what Jon Sumida has called "(Real) Absolute War."[14] Clausewitz most famously describes it as follows:

> one might wonder whether there is any truth at all in our concept of the absolute character of war were it not for the fact that with our own eyes we have seen warfare

> achieve this state of absolute perfection. After the short prelude of the French Revolution, Bonaparte brought it swiftly and ruthlessly to that point. War, in his hands, was waged without respite until the enemy succumbed, and the counterblows were struck with almost equal energy.[15]

This passage comes from Clausewitz's book on war plans (Book 8), and emphasizes his conviction that a form of absolute war could exist in reality and not just in the abstract, as evidenced by Napoleon's campaigns. Also, the topic of war plans supports a contention throughout this book that we are talking about the operational level of war—war planning constitutes a supremely *operational* process, especially as discussed in most of Clausewitz's final book, which addresses how to use military forces to accomplish objectives that have political value. However, the downside of this form of war was that it left all the most important decision-making to Napoleon. Unlike the British maritime command system that relied on collective and decentralized execution across vast distances, Napoleon concentrated all the operational decision-making for the main theater inside his own mind. As the scale of conflict grew and the talent pool of French generals able to operate independently of Napoleon declined, this unity would turn into a deadly trap—but for the first years it yielded a harvest of unprecedented success. Napoleon longed for the telegraph, and in fact created a good semaphore telegraph that gave him key advantages in 1805 as he launched his campaign. Nonetheless, his ability to communicate and direct actions in the expanding dimensions of the operational level of war—an expansion for which he to a great degree was responsible—constituted one of his Achilles' heels.[16]

ULM-AUSTERLITZ: THE TRIUMPH OF THE SEAMLESS CAMPAIGN

Recall that the Soviet operational artists desired a campaign of seamless, sequential blows. Whether from a standing start or after a period of defensive activity, they preached the merits of an approach that might avoid the "culminating point of victory" that had seemed to bedevil all sides in World War I.[17] Operational art also involves distributed and deep operations on a broad front, reflecting not so much mechanical stops and starts, but rather fluid dynamics, the motions and eddies of fluids interacting. Napoleon's first advantage in 1805 was his enemies' phenomenal failure to concentrate their forces against him in the decisive theater. The Third Coalition included the Kingdoms of Sweden and Naples and accordingly the allies decided to leverage their advantage in manpower in much the same way they had in 1799, by trying to accomplish all their political objectives with associated military efforts in all the various theaters of war. After August 1805, when the Coalition took its final form, it had at its disposal over 500,000 Austrian, Russian, British, Neapolitan, and

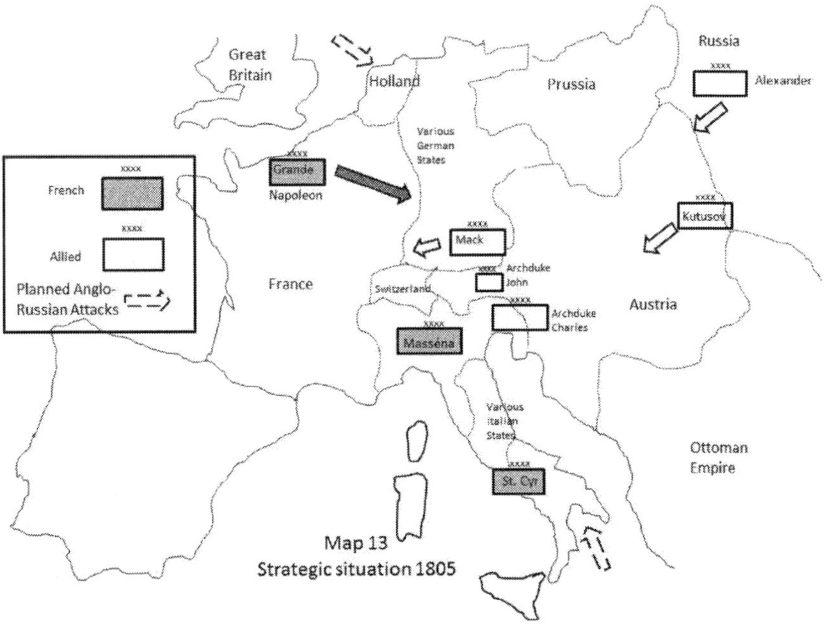

Map 13
Strategic situation 1805

Swedish troops with the very real likelihood that a neutral Prussia might join the coalition under the right circumstances.[18]

The Third Coalition's plan aimed at nothing less than the restoration of French borders to those of 1791. It intended to do this by applying all its military means to each territorial reacquisition required by this end-state. Accordingly, the Coalition's leaders dispersed their efforts on a very broad scale in order to accomplish "liberations" of Holland, Hanover, Switzerland, Italy, South Germany, and Belgium. The key was the Russian mobilization and the hope that Prussia would at least allow a Russian army (under Baron Levin Bennigsen) to pass through its territory to threaten the middle Rhine. An Anglo-Swedish-Austrian army, again passing through Prussia, was to reconquer French-occupied Hanover and from there proceed to Holland. Two other Russian armies were also on the march to join the Austrians in southern Germany—the first under Prince Mikhail Kutusov and a second army some distance behind accompanied in person by Tsar Alexander. With little concern about the distance the Russians had to march; the allies approved the Austrian plan of an immediate offensive by an Austrian army under Archduke Ferdinand and General Karl Mack into Bavaria and south Germany. Additionally, since Napoleon had twice made major efforts in Italy, Archduke Charles assumed command of the largest Austrian army in North Italy. A British, Austrian, and Neapolitan army moving up from the south supported Charles's operations and could cooperate in the expulsion of the French from the entirety of Italy. Another Russian force would operate along the Adriatic coast in support of the Austrians. Finally, a smaller Austrian army under Archduke John defended the Tyrol and threatened Switzerland. It was an ambitious program that looked much like the Second Coalition's approach in 1799 (see map 13). [19]

Napoleon made mincemeat of these plans by his blitzkrieg-like descent on the overextended Austrian army of Archduke Ferdinand after it had invaded France's ally Bavaria. Ferdinand and his chief of staff *Feldmarshalleutnant* Karl Mack halted in an exposed position in Swabia west of Munich, awaiting the reaction of the French, as well as the arrival of the Russians, rather than continuing the offensive. Also, they operated in the belief that theirs was not the major theater and that Archduke Charles, with the bulk of the Austrian army, would absorb Napoleon's main blows in Italy just as it had in 1796 and 1800. The French would be pinned down in a bloody slugfest as the Anglo-Russian-Neapolitan Army came up from the south to threaten their flank.[20]

Napoleon, however, needed a decision quickly—before Prussia removed itself from its metaphorical fence perch. This meant he must strike down the Danube at the sources of Austrian power, destroying large portions of the Austrian army and whatever Russian forces that might be in the vicinity. The Austrians' ill-advised offensive into Bavaria played right into Napoleon's hands, as he would be succoring an ally who had been violated by Austria, adding its not inconsequential military (25,000 troops) to his own on good terms. This approach also secured Napoleon's lines of communications down the Danube through southern Germany and back to France through friendly country.[21]

Napoleon's use of his many army corps and cavalry reserve to this end are justly famous. Murat was sent with the cavalry reserve to form an operational screen for Napoleon's broad advance along a front of several hundred kilometers from the northwest across the middle of Rhine. Murat's force not only covered this advance, but it also distracted the Austrians' attention toward the Black Forest and Baden instead of upon the storm moving rapidly across their right flank. (see map 14) Mack and Ferdinand believed the deception and remained fixated on the threat to their west, which they thought was advancing through the narrow passes in the Black Forest. Once convinced his operational deception had succeeded, Murat doubled back and used the mobility of his cavalry to appear on Mack's rear and flank. Murat also took over the containment of the Austrian forces along the north bank of the Danube. This operation remains a text book example of operational deception and economy of force. Meanwhile, French troops marched rapidly to the Danube and began crossing on September 25.[22]

From Albeck outside Ulm all the way to Ingolstadt on the Danube almost 100 miles away, French corps developed, fixed, or scattered Austrian detachments and concentrations, arriving deep in the Austrians' rear at Munich by October 6. Meanwhile, the Russians under Prince Kutuzov, one of Suvorov's ambitious protégés, plodded slowly up the Danube, unaware that the Austrian army at Ulm was at risk. By the time Ferdinand and Mack realized the danger, it was too late. Attempts to break out to the north failed—although they considerably disrupted French logistics. One French division fought off an entire Austrian corps on October 11 at Albeck (just north of Ulm), but was rescued by the arrival of its parent corps (Marshal Michel Ney's VI) the next day. Ferdinand and about 6,000 Austrian cavalry broke out of the trap, abandoning the hapless chief of staff, "the unhappy Mack" and half of his army inside Ulm. By bluff and demonstrations of force Napoleon hustled Mack into surrendering over 27,000 Austrian troops at Ulm on October 21, ironically the same day Nelson destroyed the Franco-Spanish fleet at Trafalgar. In the aftermath of the capitulation at Ulm, another 25,000 Austrians fell into French hands, essentially eliminating an Austrian field army by the use of maneuver and effective, if unintended, engagements. Napoleon had issued his troops two pairs of boots, and it was probably around this time that they discarded the first pair, if not soon afterward. They famously joked that their emperor won his battles with their feet![23]

Ulm emphasized all the major elements of Napoleon's operational art. His goal of finding and fixing his opponent was achieved when he managed to place the *Grand Armée* east of Ulm across Ferdinand's line of retreat. Further, he had concentrated the bulk of his army on the very spot that the Austrians and Russians needed to concentrate their own forces. Although Napoleon was not quite sure of the location of the main Austrian army, he did know that he had placed himself on interior lines to prevent its concentration. As a result, his subordinates defeated Austrian attempts, both to concentrate and to escape, in detail. In the process he bagged the bulk of the Austrian artillery and infantry. Mack neither had the imagination nor the fortitude to break out once he realized his predicament—although Ferdinand did manage to get away with about half of the cavalry he had broken out with.

At the same time Ulm was surrendering, the situation was becoming more complicated. Masséna performed his role of fixing Archduke Charles's larger army by attacking at Caldiero on October 29 and being repulsed. Charles deemed himself unable to send reinforcements through the Tyrol to help Mack. The entire Bavarian Army was incorporated into the *Grand Armée* and put to use immediately in the corps of Marshals Bernadotte (I) and Ney (VI). Meanwhile, the VII Corps of Marshal Pierre Augereau arrived from its station in Brittany and fulfilled the function of a second operational echelon, scooping up wayward Austrian detachments as it moved east to support Napoleon. On the negative side, the Prussians were beginning to mobilize after French violation of their principality of Anhalt (by Bernadotte) and

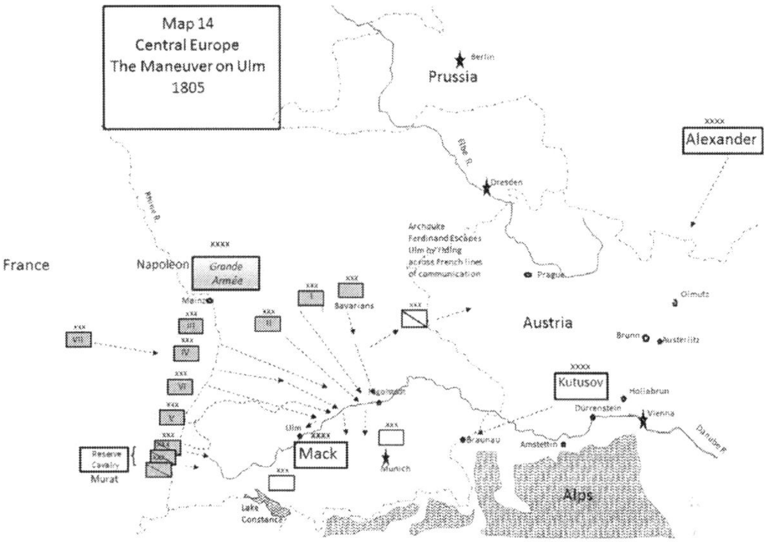

it was clear to the Austrians and Russians that they must avoid further battle. The first part of the campaign was over, but not the war. If Prussia came into the conflict, Napoleon's strategy of concentrating the bulk of his combat power into one theater might backfire on him, despite the auspicious beginning. Also, he had to worry about Archdukes Charles and John combining and abandoning the defense of the southern Hapsburg realms and advancing on his flank (and later his rear).[24]

Napoleon now tried to reap further rewards from his success at Ulm by destroying the exposed Russian army under Kutusov, which had advanced as far as Braunau on the Danube (see map 14). Kutusov fell back as soon as he learned of Ulm. Clausewitz has pointed out that the fruits of victory are often squandered when generals fail to follow up victory by a vigorous pursuit. Napoleon can hardly be criticized in late October 1805 on this account; his forces were scattered and in a state of disorganization with many dangers still to face, yet he pressed on with the easternmost corps, hoping for the best. Kutusov frustrated this purpose by deftly playing a game of hide-and-seek along the Danube, which forced Napoleon to advance down both sides of the river. The French army started to dwindle in size as it bled off troops and combat power to secure its lines of communication and as new missions were given to others to account for the Austrian forces in the Tyrol. After a sharp engagement at Amstettin with the advanced guard under Murat, Kutusov put the river between himself and the French (see map 15).[25]

In response, Napoleon created an ad hoc corps under General Edouard Mortier from several divisions belonging to other corps for operations along the north bank of the Danube. This corps became isolated north of the Danube and offered Kutusov the chance to destroy it at Dürrenstein on November 11. A tenacious defense and the operational durability of a French division under General Gazan served to keep the Russians at bay until the division of General Dupont came up, causing the Russians to break off the action. On the south side of the Danube, Davout's III Corps was able to scoop up the bulk of Merveldt's Corps on November 8. However, the damage was done and Kutusov retreated rapidly toward a second Russian army of 40,000 men, led by General Buxhowden and the Tsar, then located in Moravia.[26]

Napoleon now improvised a new plan of operations to trap Kutusov before he could join up with Buxhowden. He ordered Bernadotte's relatively fresh corps north of the Danube while informing Murat that if he could capture one of the bridges over the river in Vienna, he was to push north with his cavalry and Marshal Jean Lannes's V Corps to take Kutusov in the flank or rear as he was pursued by Bernadotte and Mortier. The first part of this plan went awry when Bernadotte expended more than two days to cross the river and begin his pursuit. As it turned out, Murat and Lannes captured the bridge in Vienna with no losses but in pushing north against Kutusov they ran into his strong rearguard at Hollabrunn under the redoubtable General Peter Bagration on November 15. Here Kutusov tricked Murat into agreeing to an armistice, allowing the bulk of his army to escape and join forces with the Tsar on the road to Olmutz. Napoleon was furious when he found out about the armistice and ordered Murat to attack immediately. Bagration conducted his usual ferocious rearguard action but lost 2,000 men before slipping away. By November 23, Napoleon and the lead elements of the *Grande Armée* were in Brünn.

It is worth reviewing just where the various components of the *Grande Armée* were at this time. A glance at the map reveals that Napoleon's forces were spread across

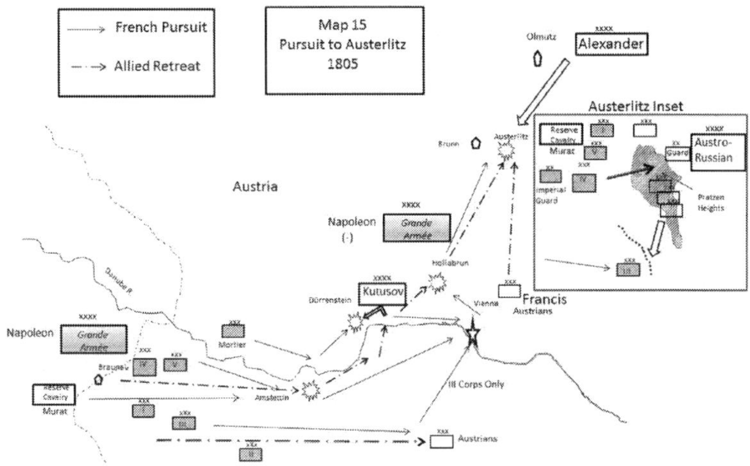

most of central Europe, from Bavaria to the Dalmatian city of Ragusa (modern-day Dubrovnik) and from Tuscany to Brünn. By the end of the month, the emperor was near the little village of Austerlitz in Moravia, with the corps of Soult (IV) and Lannes, Murat and the bulk of the cavalry, the grenadier division of General Oudinot, and his Imperial Guard. Both the corps of Davout and Bernadotte were over a day's march away. Napoleon now made one of the most effective, and short, retreats in his long career. He had occupied Austerlitz and the high ground just west of it, the Pratzen Heights. When the now unified and reinforced Austro-Russian army of 90,000 men advanced from Olmutz, Napoleon deftly pulled back from Austerlitz and abandoned the high ground on November 30, retreating just beyond the low hills to the west, which hid his forces. His new line was positioned on reverse slopes, somewhat in the manner that Wellington would employ in Spain, anchored by the Santon Hill in the north and various strong points in the south around the castle of Sokolonitz and the village of Telnitz. These last two locations were lightly held by detached elements from Soult's corps. The remainder of the army was hidden from view by the aforementioned low hills. His pickets and cavalry aggressively denied the Allies the ability to determine just where his forces were (see map 15 inset).[27]

The Allied army, learning of Napoleon's withdrawal, occupied the high ground on December 1 in the confidence that Napoleon was preparing to withdraw rapidly toward Vienna. Napoleon's retrograde maneuver also convinced the Tsar and the coterie of young and impetuous officers surrounding him that Napoleon was fearful of an attack—so they should attack, immediately. Kutusov, nominally in overall command, was against any sort of offensive and favored letting Napoleon withdraw unmolested for the time being. The Austrian staff officer Weyrother presented the Tsar a plan, without Kutuzov's endorsement, in which the bulk of the Allied army, over 50,000 men, would debauch upon Napoleon's apparently overextended and weak right flank, and cut him off from a retreat upon Vienna and reinforcements. One can hardly imagine Weyrother doing something like this when he had served under Suvorov—but Kutusov and Weyrother had no such relationship. Further, the Allies would then pursue whatever remained of the *Grande Armée* into Bohemia where presumably the Prussians would finally make their appearance as a full-fledged member of the coalition. Here Napoleon could be finished off once and for all.[28]

Napoleon intended this response. He knew he was too weak to attack the Allied Army and his parleys prior to the battle with the representatives of the Tsar and Austrian Emperor convinced them in turn that he was in a vulnerable predicament. However, by December 1, Bernadotte's corps had arrived and was in position on Napoleon's left flank to further deceive the Allies about the weakness of Napoleon's right. He was banking that Davout, marching hard from the vicinity of Vienna, would arrive en echelon during the battle to meet and fix the Allies' main attack on the presumably vulnerable right flank. This was a supremely operational calculation, based on Davout's demonstrated competence and Napoleon's judgment about how the Allied attack would develop in the short winter day of early December. Napoleon was convinced that the Allies would denude the strongest point in their line, the Pratzen Heights, leaving themselves open to a devastating counterattack. Also, like Nelson at Trafalgar, Napoleon ensured all his marshals and many of his soldiers knew the outlines of his plan and intentions for them. He visited campfire after campfire on the cold winter night of December 1–2 after a meal with his marshals and staff to encourage and explain things to his *grognards* ("grumblers").[29]

The resulting Battle of Austerlitz on December 2 is justly famous as Napoleon's tactical masterpiece. However, Napoleon's *operational* combinations *prior to the battle*, much as Nelson's prior to Trafalgar, did as much to determine the outcome as the actual fighting. His only major decision during the battle involved assessing the moment when the Allied forces attacking his weak right flank had become sufficiently vulnerable for him to launch the larger part of Soult's Corps to retake the Pratzen Heights and hit the Allied columns in the flank and rear. He did this after the bulk of the huge Allied column had proceeded south. Nonetheless, hard fighting took place across the front but with a French corps now on the high ground separating the two unequal wings of the Allied Army. Napoleon had placed himself again in the key central position. The Allies fought furiously, but the disruption and shock of this action unhinged their leadership and dispirited the troops. The Russian troops broke after their final counterattacks uphill against the French failed, mostly due to the deft use of artillery by the emperor's gunners. As they fled across the thinly frozen lakes and ponds of south of the Pratzen, the French artillery broke the ice and caused further casualties from drowning and exposure in the frigid waters. By the end of the day the French had killed or wounded over 14,000 enemy troops and captured another 11,000.[30]

The catastrophe of Austerlitz proved too much for the Austrian emperor Francis, who agreed to an armistice subsequent to further negotiations for a permanent, but disadvantageous peace. The Russian tsar, humiliated that his first battle had been a resounding defeat, in no small part because of his poor judgment, retreated toward Russian Poland with a much reduced Russian army (mostly composed of the seemingly indestructible Bagration's corps). The Prussians, who had mobilized their army and were on the point of attacking Napoleon's strategic flank and rear, hastily demobilized and sent out messages of peace and conciliation to Napoleon. The Swedes and British in Pomerania had no way to attack the French Empire, at least by land, and shelved their plans for an invasion of Hanover and Holland. Although the aftermath of Austerlitz marked the end of the Third Coalition, the Russians, British, and Swedes were still technically at war with Napoleon. Still, it seemed for a time that Europe might at last enjoy a period of peace. As for Naples, it was overrun in early

1806 and the Napoleonic Kingdom of Naples established with Napoleon's oldest brother Joseph as King.[31]

THE JENA CAMPAIGN—NAPOLEONIC OPERATIONAL ART À OUTRANCE[32]

The anti-French government of William Pitt the Younger, its architect mortally ill, was replaced by the more conciliatory government of Charles James Fox. However, peace feelers between the British and French ended up contributing to another general conflagration and central Europe was given no respite from war. Trafalgar had changed the strategic calculus and Napoleon knew it. He could not reasonably entertain the notion that he might ever be able to invade Great Britain and he only had two real options to bring about peace. One was the stick he had been using all along, to close more of Europe's ports, such as those of Austria, now an unwilling French ally, to British shipping. The other was a carrot—he needed to offer something tangible to the British for a favorable peace. Napoleon thought the electorate of Hanover, German origin of the reigning British royal family, might do the trick. However, the emperor had already promised Hanover to Prussia as the reward, or rather bribe, for staying out of the Third Coalition. When Prussia's leaders learned of Napoleon's duplicity, the war faction led by the Prussian King's beautiful and Franco-phobic queen Louise clamored for war with France. By August 1806, Prussia had secretly decided for war and entered Saxony that September. In a show of mindless bravado, officers of the elite Guard du Corps regiment sharpened their swords on the steps of the French embassy. On October 7, Prussia delivered an ultimatum to France that was tantamount to a declaration of war, but instead of waiting for her new Russian allies to unite with them, the heirs of the army of Frederick the Great marched off in great pomp and arrogance to humble the hubris of France.[33]

The Prussian army of 1806 has often been criticized for being an outdated military institution, living on the glories and traditions of "Old Fritz," impervious to reform, and completely unready for a war in the new modern style.[34] This was not entirely true, although it certainly proved no match for Napoleon. Ever since Prussia's withdrawal from the First Coalition in 1795, significant members of Prussia's officer corps made study and knowledge of the new French operational methods of war their focus. Unfortunately, these elements, which we shall label the *reform faction*, resided inside a relatively new institution, the Quartermaster General Staff (QMGS, founded originally by Frederick the Great). This unique group of officers was analogous to the French staff officer corps of the prerevolutionary era from which officers like Berthier sprang. The staff was composed of three sections and also involved an element for officer education. After the catastrophe of 1806 and 1807, this group controlled the Prussian military and created an entirely new operational framework to oppose Napoleon and the French—the Prussian Great General Staff.

To fully understand the origins and purpose of the QMGS and new *Kriegsakademie* (War College), one must go back to Prussia's enlightened policy of acquiring talented non-Prussian officers to serve in it ranks. The Prussian military institution had always had a very close relationship with the Dukes of Brunswick, sovereign German princes in their own right. Prussia often used them as independent commanders, giving them field marshal rank. Charles William Ferdinand, Duke of Brunswick (defeated by Kellermann at Valmy), came from the old school of generalship but was an astute judge of men. It was on his recommendation that the founder of the new *Kriegsakademie* and head of the third section of the QMGS, a Hanoverian officer named Gerhard Scharnhorst, came into the Prussian army.[35]

Scharnhorst came from a humble background; his father was a soldier but not an officer, but did well enough to earn a place for his son in the Hanoverian cadet school. While serving as chief of staff for the Hanoverian army, Scharnhorst's combat record and writing brought him to the attention of the Duke of Brunswick. Brunswick recommended him to the King of Prussia and Scharnhorst transferred into the Prussian service as a major, and shortly after, the King promoted him to lieutenant colonel.[36] He served simultaneously in two jobs. His first and more influential position was as the director of the revamped *Kriegsakademie* that trained infantry and cavalry officers. Here Scharnhorst met the young Clausewitz and formed a close professional friendship that lasted until the former's death in 1813. Additionally, Scharnhorst was appointed to the QMGS that was reorganized by Colonel von Massenbach in 1803 into three logistics planning sections. Scharnhorst headed up the section assigned to contingency planning for operations in the west against the French. According to Clausewitz, Scharnhorst was "one of those rare individuals in whom theoretical knowledge and scholarly ambition are combined." Scharnhorst used these skills to educate his students and his planning section on what he correctly viewed as the profound changes in war wrought by the French Revolutionary and Napoleonic Wars. However, the old guard in the Prussian army remained aloof from new ideas and resistant to the idea of military reform.[37]

Scharnhorst and his confederates, men such as Clausewitz, Gebhard von Blücher, Baron Carl von Müffling, and to a lesser degree Brunswick, attempted to reorganize the Prussian Army on the eve of war with France. They were opposed by the bulk of the officer corps, which included over 142 generals, half of whom were 60 years or older, including four older than 80 years. The Prussian army, like the Royal French army, was mostly officered by aristocrats who had inherited and had not earned their commissions. Despite some improvements in skirmishing procedures, the Army retained the linear tactics of Frederick the Great. The Prussians did try to organize combined arms divisions at the eleventh hour in 1806 after the field army marched toward the seat of war in Saxony, but this perhaps hurt more than it helped. There was plentiful artillery, however, it was scattered about with no concept of how to use it as mobile fire power to weight a battle. The cavalry was superbly mounted, but had little experience in combined arms warfare and was not organized to screen and provide reconnaissance for the army. The cavalry's self-image was that of the most important arm, deciding the battle much as Baron von Seydlitz had at the Battle of Rossbach (1757) by charging "hell for leather" at its enemies. Finally, there was no standardization of organization above the division level, no army corps counterpart to the French. The army was visually impressive, but it was an antique, with every bit as much baggage as a standard 18th-Century army but without Frederick's genius to move and maneuver such a cumbersome machine, either operationally or tactically. Clausewitz remarked of it, "behind the fine façade all was mildewed."[38]

The greatest mistake that the Prussians made, after declaring war in the first place, occurred at the strategic level. Ignoring the cautious advice of Scharnhorst to wait for the Russians, Brunswick and Prince Friederich Hohenlöhe, who commanded the next largest army, spent precious days arguing over which axis to advance from Saxony while Napoleon rapidly concentrated his forces as he had done in 1805 for the main effort in the decisive theater. The Prussians launched an offensive toward the

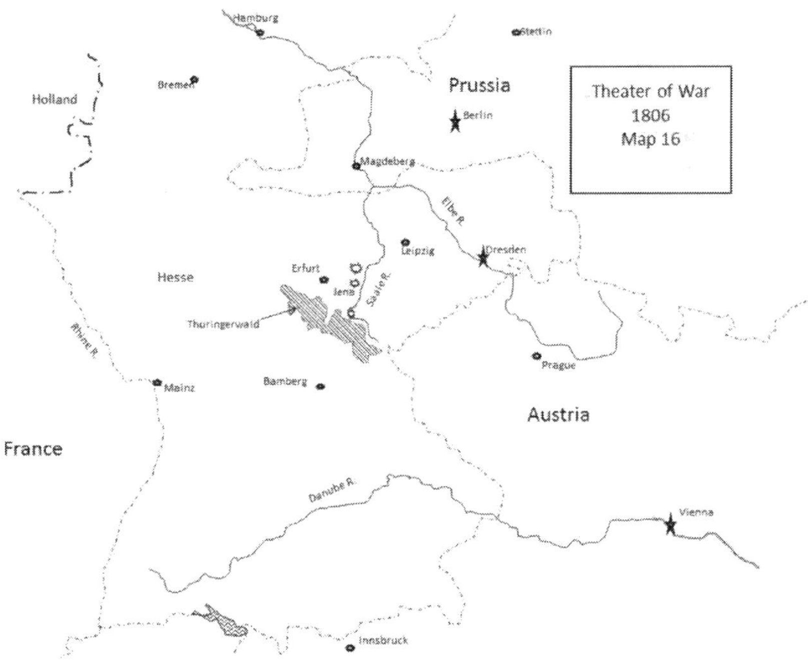

eastern side of the Rhine before they had completely mobilized or given the Russians time to join them. They and their unwilling Saxon ally simply grouped themselves into three armies of varying size: the largest numbered about 70,000 troops and commanded by Brunswick who also served as overall commander in chief, Hohenlöhe's army (50,000), and General Philip von Ruchel's small army (15,000). They advanced with fewer than 150,000 troops as opposed to Napoleon's battle-tested, rested, and victorious *Grand Armée* of over 180,000. There were additional fortress garrisons and detachments that the Prussians had elsewhere, but in moving out to meet Napoleon they did not bring all their available combat power with them, leaving a "strategic" reserve of 13,000 out of supporting distance to their rear.[39] Writing not long after participating actively in this campaign, Clausewitz claimed desertion and straggling lowered the effective Prussian-Saxon field force to approximately 130,000 effectives.[40]

For once, Napoleon's strategic intelligence agents failed him and only after the Prussians had entered Saxony did he realize that war was imminent. His hopes for peace treaties with Russia and Great Britain proved vain and he concentrated the bulk of his combat power against the Prussians. The *Grande Armée* was scattered from the Inn River to its northernmost cantonments in Hesse along the lower Rhine (see map 16). By September 18, Napoleon had made up his mind and a flurry of activity ensued at his headquarters. He issued an amazing 120 separate orders whose intent was to concentrate the bulk of the *Grande Armée* near Bamberg just west of the dense Thuringerwald Mountains through which he intended to advance into Saxony. The Imperial Guard infantry in Paris were loaded into wagons and moved rapidly along good roads to Mainz on the Rhine where they disembarked and continued on foot to the east. Napoleon and his staff followed a week later. By the first week of October, the bulk of his forces were concentrated between Bamberg and Wurzburg, before Prussia's ultimatum arrived on October 7.[41] Earlier, Napoleon had also issued orders tantamount to the mobilization of a second strategic echelon (using the Soviet terminology) by calling up 50,000 conscripts of the class of 1806 as well as activating another 30,000 French reservists (troops who had been discharged after earlier campaigns but were still on the rolls of the French army). In this way he mobilized the necessary reserves to either reinforce his army for a longer effort to the east against the Russians or take defensive measures. He also learned the Russians were joining Prussia, Sweden, and Britain in a Fourth Coalition.[42]

Napoleon decided to advance through the three narrow passes of the Thuringerwald, thus screening his army. He sent his infantry forward first in order to get the maximum amount of combat power through the dense hills. He counted on the Prussians not reacting swiftly enough to defeat the heads of his columns as they emerged from the passes as well as on the audacity and speed of his own advance. His campaign design is best gained from a complete reading of his famous operations order sent to Marshal Soult after Napoleon arrived at Wurzburg:

> I propose to debouch into Saxony with my whole army in three columns. You are at the head of the right-hand column with Marshal Ney's Corps half a day's march behind you and 10,000 Bavarians a day's march behind him, making altogether more than 50,000 men. Marshal Bernadotte leads the center, followed by Marshal Davout's Corps, the greater part of the reserve Cavalry, and the Guard, making more than 70,000 men. He will march by way of Kronach, Lebenstein and Schleiz. The V Corps [Lannes] is at the head of the way to Coburg, Grafenthal and Saalfeld, and musters over 40,000 men.
>
> With this immense superiority of force united in so small a place you will realize that I am determined to leave nothing to chance, and can attack the foe wherever he chooses to stand with nearly double his strength. . . . If the enemy opposes you with a force not exceeding 30,000 men, you should concert with Marshal Ney and attack him. . . . From news that has come in today [October 5] it appears that if the foe makes any move it will be towards my left; the bulk of his forces seem to be near Erfurt. . . . I cannot press you too earnestly to write to me frequently and keep me fully informed of all you learn from the direction of Dresden. You may well think that it will be a fine thing to move around this area in a "battalion square" of 200,000 men. Still, this will require a little [operational] art and certain events.[43]

Not mentioned in this order was Augereau's VII Corps, which followed in the wake of Lannes (see map 17). Napoleon had organized his army into a giant square composed of roughly six army corps. His method here differs somewhat from that employed at Ulm in that this operational scheme gave him the maximum amount of flexibility to initiate one of his operational meeting engagements. Whichever front

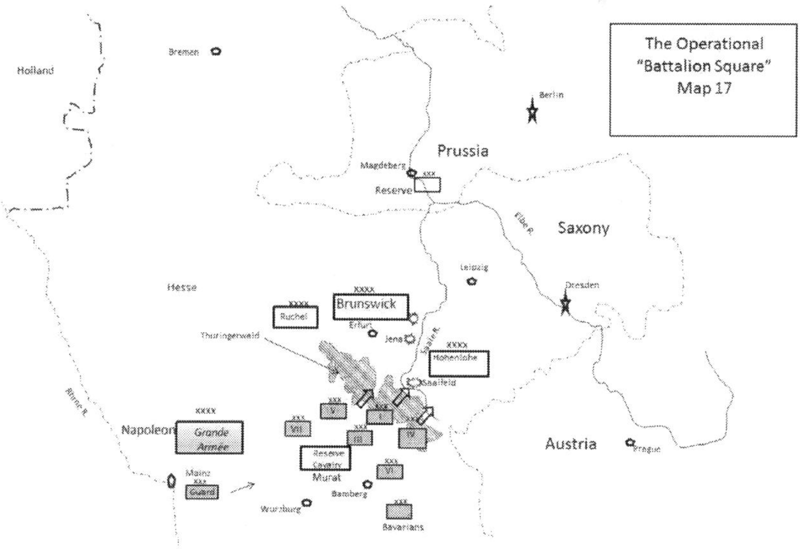

met the enemy would fix it and then engage it decisively in concert with the second line of operational reserves. At first there would be one axis determined absolutely by the passes through the mountains, but once into the more open country beyond in the Saale River valley, Napoleon would in fact turn the entire axis from the northeast to a more northerly direction toward Erfurt. In addition to the second line of operational reserves, in some cases he even had a third line, such as the large Bavarian division under Napoleon's brother Prince Jerome.

The Prussians and their archaic style of command played right into Napoleon's operational scheme. A reconnaissance by Scharnhorst's protégé Captain Müffling revealed that Napoleon had already left Bamberg with a very large force marching for Saxony. Previously, Brunswick and Hohenlöhe had argued about what to do, whether to threaten Napoleon's rear and flank by advancing further north or push on and meet him east of the Saale River. Brunswick dithered and cancelled his moves from Erfurt toward Wurzburg. Hohenlöhe interpreted this to mean that this allowed him to execute his alternate plan of moving on Bamberg along the eastern side of the Saale. Hohenlöhe pushed out his strong advanced guard division under the King's cousin, the firebrand Prince Louis Ferdinand of Prussia, to Saalfeld and another division that included 3,000 sullen Saxons under General B. F. Tauentzien to Schleiz.[44]

Referring to Napoleon's order to Soult, these rash moves placed Hohenlöhe's forces right in the path of Lannes's left-side column and Bernadotte's strong central column, both now already through the *Thuringerwald* (Thuringian Forest). Disaster struck quickly. Bernadotte pushed Tauentzien out of Schleiz on October 9. Although casualties were low, Tauentzien's division was thoroughly disrupted and was lucky not to lose its Saxon troops completely to capture or defection. Hohenlöhe pulled them back to reorganize them. Louis's division was then mauled by Lannes's infantry and cavalry on October 10 at Saalfeld and the prince was killed in personal combat by a French hussar. Here Prussian losses were significant, totaling 2,700 as opposed to less than 200 for the French. These initial defeats unhinged Hohenlöhe and, especially, Brunswick, who now ordered his armies to concentrate northwest of the Saale River around Weimar. Napoleon interpreted Brunswick's moves to indicate he was pulling back east of the river to protect Leipzig, the capital of Saxony. In doing so, Napoleon "lost" the Prussian army, as he had the Austrians at Ulm. Napoleon's giant operational square, which had been pivoting to the north, now turned clockwise to the northeast.[45]

As the Prussians withdrew, French cavalry further discomfited Hohenlöhe's Saxon troops. Intelligence arrived at Prussian headquarters that Napoleon already had French troops (in fact Davout's entire corps) as far north as Naumburg, which threatened the Prussian line of retreat. The fog of war had set in deeply for both sides, but Napoleon had the better tool to hand to deal with uncertainty. On October 13, Lannes, now presumably in the second operational echelon, discovered Hohenlöhe's entire army of about 40,000 men on the flank of the *Grande Armée* at Jena. Using dragoons as infantry, Lannes seized the key terrain of the Landgrafenberg Heights near Jena in the immediate vicinity of Hohenlöhe (perhaps not realizing how strong he was). He then notified Napoleon of his situation, an example of the success of Napoleon's style of mission command, best illustrated in this campaign, with trusted subordinates like Lannes and Davout. Napoleon's plan at this point now changed as he realized the enemy was not withdrawing to cover Leipzig but seemed to be across the Saale at Erfurt. As information came in, he conceived a plan to advance again at 90 degrees to his previous axis (toward Leipzig) and on Erfurt where he now believed the main enemy army to be, anticipating a large battle on October 16.[46]

Lannes's updated information about the actual size of the Prussian army at Jena (he estimated 30,000) set the stage for the next day's cataclysmic events. Napoleon, believing that the Prussian main army was at Jena, issued a stream of orders to bring the main force to bear there, increasing from 22,000 to over 130,000 troops at Jena as the day wore on. It was Suvorov's echeloned arrival on steroids! Napoleon still did not understand Prussian intentions. Once they learned that elements of the French army were as far north as Naumberg, a Prussian council of war decided to retreat to Magdeburg down the western side of the Saale through Merseburg. Hohenlöhe would cover this withdrawal with his army of approximately 38,000 at Jena and could call on Ruchel's small army, moving across his rear in the wake of the main army, if he needed support. Brunswick, meanwhile, would take the main army just west of Naumberg through Auerstadt on the way to Freiburg and Merseburg. Thus, as October 14 dawned, Napoleon was moving upon Hohenlöhe's sizable rearguard with the bulk of the *Grande Armée*.[47]

Even though he misunderstood the enemy's intentions, Napoleon's forces were well positioned to deal with the planned Prussian movements. Murat and Bernadotte had been given orders to concentrate at Dornburg, midway between Auerstadt and Jena, and could support Napoleon or Davout as events dictated. Davout was to advance south toward Auerstadt and take the Prussians in the flank or rear and to bring Bernadotte under his orders if combat ensued. Bernadotte had the same orders but interpreted them to mean that he need not support Davout, while Murat pressed south to join the Emperor at Jena. In this manner, Davout and his single corps of 26,000 troops marched directly into the path of the retreating Prussian army under Brunswick of almost 64,000 troops[48] (see map 18).

October 14 became the blackest day in the Prussian army's history. Napoleon arrived at Jena to find that Marshal Ney had launched the offensive there prematurely, and had been repulsed handily. Nonetheless, as his operational reserves arrived, the emperor fed them into the battle and before long the *Grande Armée* had shattered Hohenlöhe's army. Ruchel moved up too late in support and was in turn mauled by the victorious French who used massed artillery consisting of four batteries of guns

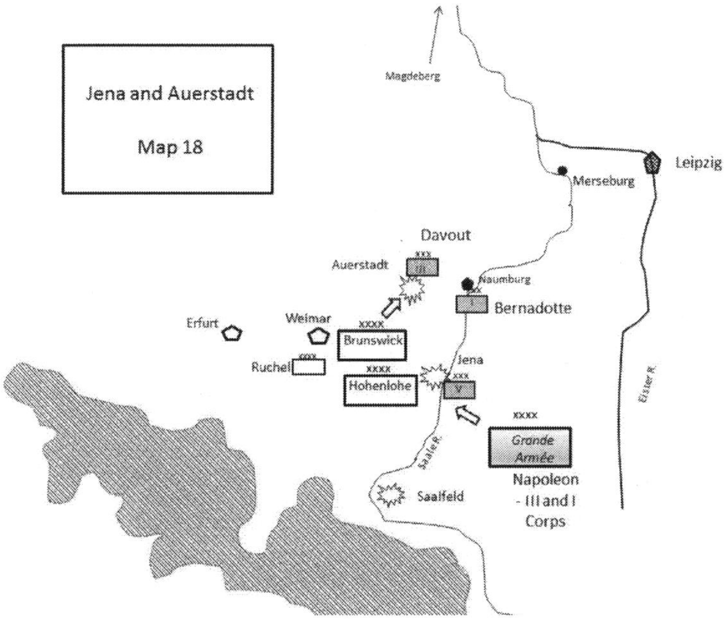

to unnerve his infantry and then scatter them. Napoleon believed that he had shattered the main Prussian army only to receive a dispatch late that night from Davout, over 20 miles away at Auerstadt, informing him that his lone corps had successfully repelled the main Prussian army under Brunswick. Brunswick had been mortally wounded, leaving the Prussian army leaderless at its most critical moment. Davout's III Corps had then resumed its advance and the entire Prussian army literally disintegrated on the field of battle. Bernadotte had never supported him, despite Davout's urgent need and the furious sound of cannon just to his north. In contrast, Soult marched to the sound of the guns, despite not receiving any definitive orders from Napoleon, getting his lead division into the fight at a critical moment on the enemy's right flank at Jena. Davout's excellent generalship aside, his corps's performance at Auerstadt stands as a testament to the durability of Napoleon's *Corps d'Armée*. French losses for both battles totaled 12,000 (mostly in Davout's and Ney's corps) while Prussian losses exceeded 39,000 killed, wounded, and captured, with numerous units fleeing as disorganized refugees and the army's line of retreat cut.[49]

The campaign was not over. Clausewitz used Jean-Auerstadt as the supreme example of how a pursuit after a hard-won victory (or two in this case) increases exponentially the scale of an enemy's defeat if conducted with the proper energy and boldness.[50] Before light on October 15, Napoleon began to issue orders for an operational pursuit. Napoleon had his reserve cavalry corps, Bernadotte's I Corps (due to that marshal's indolence), and most of Soult's Corps relatively fresh and in hand. They ended up serving as essentially a second operational echelon, poised to gather up the fruits of victory operating as what the Soviets called the deep exploitation force. Napoleon unleashed these against the fleeing and disorganized Prussians, giving them no chance to reform and establish a defense before Berlin. He planned to send Murat, Ney, and Soult toward Erfurt and then turn them north toward the Elbe to apply pressure on the mass of refugees that the Prussian army had become. On this force's right was the intended maneuver force of decision, four corps led by Bernadotte's fresh troops, which Napoleon intended to reach the Elbe before the Prussians. The postbattle pursuit by Murat, using mostly the combined corps' cavalry at Jena, yielded its first substantial success that day when the column of the Duke of Saxe-Weimar (another "detached" Prussian contingent) was bluffed out of the fortress of Erfurt and Murat gathered up some 9,000 additional prisoners from units fleeing along that axis by the end of October 16.[51]

On the Prussian side the goal was to reach Magdeburg and establish a defense along the Elbe before the French arrived. Three corps had formed up after October 14 from the residue of the Prussian army, the remainder of the troops from Jena under Hohenlöhe, those fleeing Auerstadt under General von Kalckreuth (20,000), and finally a third group under General Blücher, which was supposed to be serving as a rearguard for the other two. After the capture of Erfurt, Blücher joined up with Saxe-Weimar's remnants, made the wounded General von Scharnhorst his chief of staff, and retreated on a divergent axis to west of the others as advised by Scharnhorst.[52]

Bernadotte, stung by the criticism of the army and Napoleon for his conduct at Auerstadt, proceeded savagely with his portion of the pursuit. The Prussian strategic reserve under the Duke of Württemberg, numbering 13,000 fresh troops, had remained at Halle awaiting further orders. Bernadotte's corps arrived and defeated them on October 17, inflicting another 5,000 casualties against a cost of 800 men. On the other axis of the pursuit, Soult had overtaken Ney and was hot on the heels of the combined forces of Kalckreuth and Hohenlöhe. By October 20, the various French corps were either at or approaching the Elbe River. The Prussians abandoned Magdeburg and their original line of retreat toward East Prussia, reinforcements, and the Russians. Instead they veered west of Berlin, which they effectively abandoned. They were hoping the French would slow down and besiege Magdeburg, but Napoleon kept the press on, bypassing that city with Soult and Bernadotte. All the while his forces gathered up prisoners and stragglers at every turn. Ney, coming up in the second echelon, was left behind to surround and besiege Magdeburg. Blücher briefly rejoined Hohenlöhe on October 24 and assumed the role of the rearguard. The next day, Davout's corps marched into Berlin unopposed.[53]

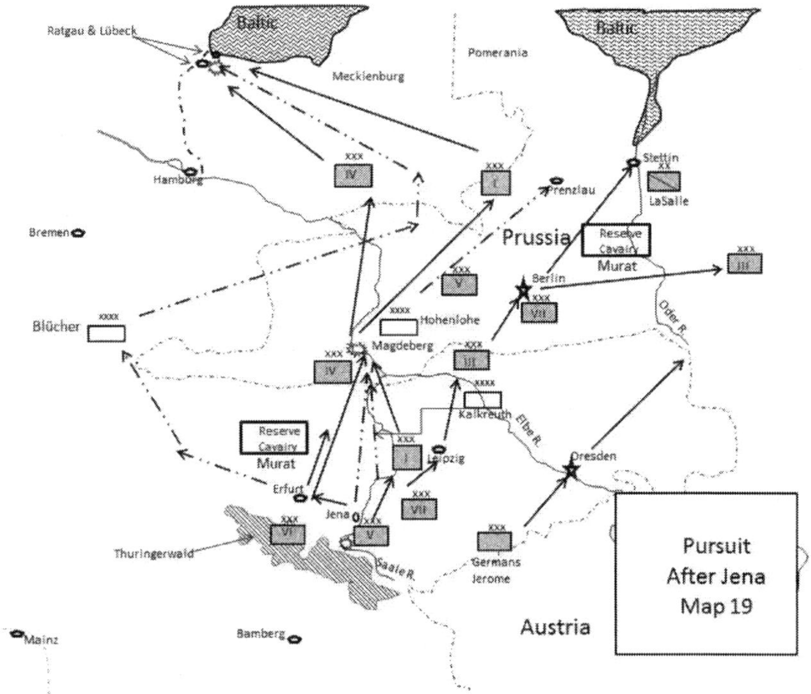

Blücher and Scharnhorst turned west and away from Hohenlöhe racing toward Lübeck, perhaps hoping to be evacuated by the British navy. Instead, they met a Swedish division that had recently disembarked. Bernadotte, still marching hard, penned up Scharnhorst in Lübeck and Blücher in Ratgau (see map 19). Under assault by the corps of Bernadotte and Soult, they surrendered with over 20,000 troops on November 4–5. Bernadotte also scooped up the Swedes, effectively removing an entire army from the operational board. He treated the Swedes so cordially that they returned to Sweden full of praise for the gracious French marshal—later these fond memories caused the Swedes to invite Bernadotte to come rule Sweden as crown prince when their chronically deranged king was removed from the throne. In 1813, he "went native," abandoned Napoleon and joined the Sixth Coalition on its march to ultimate victory. Bernadotte had taken lemons and indeed made lemonade. His dynasty still rules as constitutional monarchs in Europe today.[54]

As Bernadotte was scooping up Prussians, Swedes, and a future crown in Westphalia, Davout and Murat pursued Hohenlöhe. On October 28, Murat bluffed Massenbach and Hohenlöhe into surrendering with their 10,000 troops at Prenzlau.

Lannes's V Corps pursued the Prussians to Stettin, where his cavalry advance guard under General LaSalle, following the example of Murat, bluffed the fortress's garrison of 5,000 men to surrender to his 1600-man cavalry division. The final nail in the coffin of the Prussian army came on November 6 when General von Kleist surrendered his starving and dispirited 22,000-man garrison in Magdeburg to Marshal Ney. In a little over five weeks, the *Grand Armée* had gathered up 140,000 Prussian prisoners, mostly after October 14. It was one of the most lopsided operational victories in history.[55]

EYLAU AND FRIEDLAND

The War of the Fourth Coalition did not end when French troops entered Berlin. Despite the scale of the victory, the fact that these monumental successes had not ended the war led to muted celebrations back in France. The war-weariness seen near the end of the French Revolution was again in evidence.[56] The battles of the Polish campaigns of 1806–1807 represent both the old and the new. Ironically, the shape of things to come occurred first during the campaign that terminated at the battle of Eylau, a bloody stalemate that left Napoleon without a decisive conclusion to the war of the Fourth Coalition. At Eylau one begins to see the transforming character of the operational art of the Napoleonic War, from short decisive campaigns into those of protracted, attrition warfare. The German military historian Delbrück labeled this approach to warfare as *ermattungsstrategie*, strategy of exhaustion.[57]

Napoleon's destruction of the bulk of the Prussian army did not convince the Russians, British, or the Prussian queen, who bent her weak husband to her will, that defeat need be conceded. Napoleon's initial peace feelers were rejected by all parties. Thus, there were still large Prussian domains in Silesia, eastern Poland, and East Prussia not under Napoleon's control. He initially delayed the decision to invade Poland, but soon realized that proceeding to the line of the Vistula River had key advantages. Such a position would protect his flanks and his operations to subdue the great fortress of Danzig and secure the Prussian province of Silesia to the south. Nonetheless, he had to peel away forces to garrison the many Prussian fortresses he had seized. One small fortress at Kolberg in Pomerania—defended by an obscure Saxon in Prussian service named August Neidhart Gneisenau—remained defiant, but was masked by light French forces.[58]

Napoleon decided to continue the offensive by mid-November, when he learned that a Russian army of at least 56,000 had departed in late October to join the Prussian corps under General Anton Lestoq. The French emperor in turn detached a multinational corps composed of Germans commanded by Prince Jerome and assisted by General Dominique Vandamme to secure Silesia as he proceeded east across the Oder

River. Davout had already crossed that river on October 31 and, with the surrender of Stettin, Lannes's corps was ready to cross by the same day. Napoleon's losses had been light to this point, but nonetheless he knew hard fighting was ahead after the Prussians rejected his peace terms (November 16). Napoleon brought up the second strategic echelon he had mobilized to free his veterans for further campaigning and replace losses. He also denuded secondary theaters of cavalry units, knowing he would need much of that arm in the vast spaces of East Prussia and Poland.[59]

With the Oder breached, Napoleon's columns now advanced deep into Poland. A large component of his army (Bernadotte and Soult) had been pulled away by Blücher's detour to the west and, along with Ney's VI Corps, these forces were now several weeks away from rejoining Napoleon. He could have waited for them, but decided to push ahead and establish a defensive line along the Vistula River, between Warsaw and Thorn, before the Russians could establish themselves in strength west of that river and to forestall the Prussians from mobilizing the manpower in Pomerania and Prussian Poland to reconstitute their army (see map 20). He succeeded in this daring operational move and Murat and Davout pushed into Warsaw on November 28 as "liberators" to the cheering of Polish citizens. By late November, he had three corps and about half of Murat's cavalry along the Vistula. The Russian general Bennigsen had been bluffed out of Warsaw.[60]

The Russians and Prussians attempted a series of local counterattacks trying to catch the French off their guard but these were all repulsed throughout December as the bulk of French army now operated east and north of the Vistula. The Russians

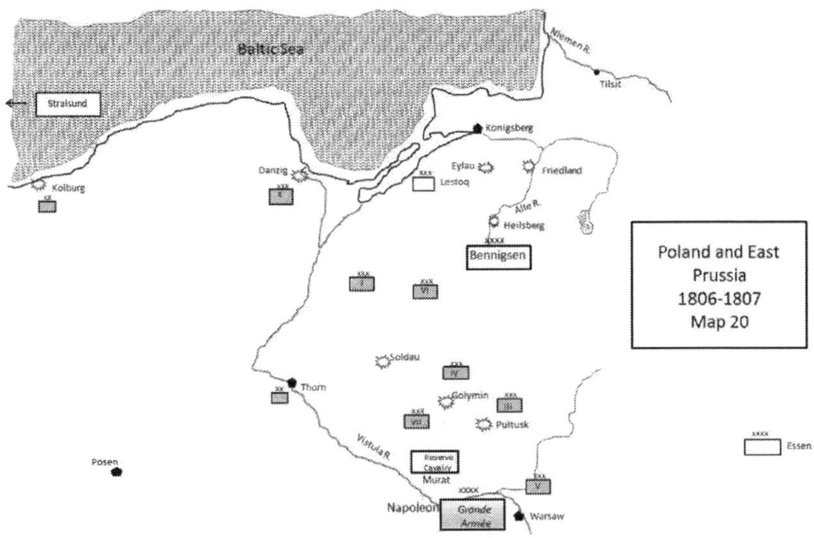

commenced a general withdrawal and Napoleon lost contact with them. The end of the month saw Napoleon regain contact and fight three corps-sized battles at Soldau (Ney), Golymin (Augereau), and Pultusk (Lannes). At Pultusk, Lannes was heavily outnumbered until one of Davout's divisions came up and the Russian counterattack failed—yet again emphasizing the operational durability of the *Corps D'Armée*. This was enough to cause Bennigsen, now in overall command of his army and Buxhowden's, to retreat. By the New Year, both armies were spread out in winter quarters throughout the frigid Polish countryside. To the northwest, Napoleon formed a new X Corps under General Victor, who was subsequently captured by Gneisenau's aggressive cavalry sorties from Kolburg. Napoleon then placed this new X Corps, mostly Germans, under the veteran Marshal Francois Lefebvre.[61]

Just as the French were settling in for the winter to cover the siege of Danzig, Bennigsen launched a powerful offensive against Ney's isolated corps. However, he merely pushed it back and then ran into Bernadotte moving up in support on Ney's flank as that marshal retreated. Bernadotte repulsed Bennigsen's advance guard and then retreated in order to stay aligned with Ney. Napoleon had many irons in the fire at this point. Jerome was fully engaged in Silesia, Lannes's corps watched to the east for the arrival of a small Russian army under General Essen, sieges continued along the lower Vistula, and Bennigsen commanded a powerful army of almost 70,000 tough tsarist troops. Nonetheless, Napoleon sensed a chance to destroy Bennigsen and he ordered an all-out offensive against the Russians' rear and left. Unfortunately, Bennigsen's Cossacks intercepted Napoleon's orders and the Russians began to retire to escape the trap. In doing so, Napoleon became strung out and this explains how he came to face Bennigsen's much larger army on the evening of February 7 just outside the little town of Eylau in East Prussia. He sent orders to Davout to join him (as he had before Austerlitz) but delayed in recalling Ney until the next morning. Bennigsen, though, recalled Lestoq's Corps at 3:30 A.M. and that officer's chief of staff, Scharnhorst, recently exchanged, plotted a deft approach march that side-stepped Ney—who was supposed to keep the Prussians away from the battle.[62]

Napoleon's haste to come to grips with the Russian army, and perhaps its past performance, led him to overreach. He had made errors in the past, after Ulm and before Jena-Auerstadt, but this was among his most glaring. His chase of the Russians had led him to end up in a cage with a "polar bear" in the extreme conditions of the Polish winter. The two-day icy holocaust of Eylau began late on February 7 and then exploded in a senseless slaughter the next day. As at Austerlitz, Napoleon was outnumbered; unlike Austerlitz, he attacked prepared enemy positions. The battle precipitated an early crisis when Marshal Augereau's small corps blundered into massed Russian artillery as it emerged out of a blinding snowstorm. Russian cannon obliterated it. To counter this disaster, Napoleon launched one of the most famous cavalry attacks in history, Murat's splendid troopers superbly mounted on captured Prussian chargers. This charge did not win the battle, but it prevented a defeat. After that, the battle became a bloody back-and-forth engagement as operational echelons from both sides arrived to weigh in on the decision. Davout's corps, the key to Napoleon's tactical plan, arrived first and this attack had just begun to bear fruit when Lestoq's 17,000 Prussians arrived on the scene to give the initiative back to the Russians.[63] Just as these measures seemed likely to break Napoleon's damaged army, Ney's corps arrived on the scene in the last hours of daylight, causing both sides to cease fighting. Ney captured the essence of the battle upon seeing the thousands of corpses in the snow, remarking "What a massacre, and without any results!"[64] Total casualties for the two days of confused fighting totaled over 35,000 and possibly higher, two-thirds of them French.[65]

Bennigsen, like Robert E. Lee after Antietam a half century later, decided that discretion was the better part of valor, and withdrew the Russo-Prussian army in good order into winter quarters. Napoleon then did likewise after half-heartedly declaring victory. Napoleon knew he had almost been bested and for the first time the luster of his generalship did not shine. In France, dissatisfaction at these seemingly unending wars began to result in serious evasions by would-be conscripts, leading to the French version of the modern draft dodger—*le réfractaire*. The losses of veterans in the *Grande Armée* had been grievous, the superb army that marched into Germany the previous Autumn and the best army of his entire career was badly damaged. More and more his army consisted of foreign elements: Poles, Italians, Saxons, Bavarians, and other contingents from the newly formed Confederation of the Rhine. Although the *Grande Armée* proved itself again in the spring and again in 1809, Eylau can be considered as the engagement that demonstrated the limitations of Napoleon's method of operational art when faced with a determined opponent unwilling to surrender and willing to trade space, time, and casualties with Napoleon.[66]

General Lefebvre and his multinational X Corps continued the siege of Danzig and its Prussian garrison. Bennigsen returned to the fray with another unexpected offensive in an attempt to break the siege. Several minor actions convinced him to return to winter quarters. Despite being spread out, Napoleon's dispositions proved resilient enough to frustrate both Bennigsen's offensive as well as two subsequent Russian and Prussian amphibious operations meant to aid Danzig in May. At the same time, a breakout by the garrison was attempted, but all these actions were ill-coordinated. The Russians were penned up in their beachhead and the Prussians repulsed. Later, the British and Russian navies evacuated the trapped Russian troops.[67]

On May 22, Danzig surrendered while Jerome and Vandamme continued to reduce Silesian fortresses. Additionally, the *Grande Armée* added more enthusiastic Polish troops to its ranks. The Russo-Prussians, instead of conserving their strength, had frittered away their small advantage after the stalemate of Eylau. Now Napoleon's flanks were secure as the spring campaign season approached; his rested and refitted army was ready for major operations. The Allies initiated things by launching yet another premature offensive against Ney's isolated corps in early June. Ney again conducted a fighting retreat that blunted the advantage Bennigsen had achieved with surprise. Lestoq, advancing in support, was also rebuffed by Bernadotte's corps. Further down the Vistula, Masséna, recalled from Naples, contained the forces of Essen on the southeast side of the Polish front. Napoleon again tried to trap Bennigsen by fixing him with the corps of Victor (who had replaced the wounded Bernadotte), Soult, Ney, and Davout (roughly aligned north to south) while advancing on Bennigsen's open left flank with his operational reserves. These consisted of four divisions of cavalry plus a new reserve corps under Lannes and Mortier's corps, which had been released from the Stralsund front after the Swedes signed an armistice. Napoleon's surplus of troops thus formed a ready-to-use second operational echelon. As usual, the cagey Bennigsen withdrew his head from the noose before it snapped shut, retreating through the woods and lakes of East Prussia toward his fortified camp at Heilsberg. Ironically, this was the same ground on which the Russians would suffer a disastrous defeat over a hundred years later at Tannenberg in 1914.[68]

After a brisk rearguard action at Guttstadt on the Alle River, which now covered the retreating Bennigsen's right flank, the Russians fell back into their camp at Heilsberg. Napoleon pursued with only two corps and Murat's cavalry. On June 10, with only part of these forces closed up, Murat and Soult attacked and were repulsed and then counterattacked by Bennigsen. Lannes arrived and stabilized things, but was also repulsed in front of the Russian works. Napoleon's goal was for these forces to fix Bennigsen, but again that general withdrew. The Russo-Prussian troops withdrew along two axes, Lestoq toward Konigsberg and Bennigsen up the right (east) bank of the Alle. Napoleon, unsure of the axis of Bennigsen's retreat, pursued with cavalry on both sides of the river, but sent the bulk of his forces due north toward Konigsberg. Bennigsen planned to recross the Alle to cover Konigsberg and join Lestoq. His forces arrived at Friedland on June 13, ejected the French cavalry there, and established pontoon bridges to better move across. Napoleon had directed Lannes to secure Friedland. Lannes arrived very early on June 14 and set up a defensive blocking position. He then notified Napoleon that he had Bennigsen's entire army facing him. This established the occasion for the battle of Friedland.[69]

Bennigsen, believing he had isolated Lannes's Corps, continued to bring his 60,000 men across the river through the narrow streets of the village, but his position was a poor one and Lannes, somewhat similar to Davout at Auerstadt, skillfully fought a nine-hour defensive battle that kept the Russians bottled up around Friedland. Just after noon Napoleon arrived in person followed by 55,000 relatively fresh troops. The long June day allowed him to fight a second battle, shattering Bennigsen's army, including the use, for the first time, of a mobile massed artillery battery under General Senarmont. Napoleon inflicted almost 20,000 casualties, almost all dead or wounded, while incurring 10,000 of his own. On June 15, Lestoq abandoned Konigsberg to Soult and Davout and joined Bennigsen's army of refugees as it streamed toward Tilsit. Tsar Alexander had had enough. On June 23, he signed an armistice with Napoleon ending hostilities, and two days later, the famous meeting on a raft at Tilsit on the Niemen River took place between the two Emperors. Alexander joined Napoleon's "Continental System" of economic warfare against Great Britain and signed a Franco-Russian defense treaty. Prussia was thoroughly humbled, her army limited to less than 45,000 men, the grand duchy of Warsaw carved out of her territory, her western possessions and those of her allies used to create a new Kingdom of Westphalia for Jerome Bonaparte, and her fortresses garrisoned with Napoleon's troops, and the Napoleonic Empire had reached its pinnacle.[70]

* * *

The short Friedland campaign proved the *Grande Armée* and Napoleon again as an unbeatable combination because they reflected the new operational way of war while the Russians remained mired in the past, organized and operating under the rules of the 18th century. Nonetheless, the tactical brilliance of Friedland masked the change in the character of Napoleonic operational art that would soon be in evidence in the last part of the decade. This proved to be the last time that Napoleon would have an asymmetrical advantage over his opponents as they adopted the French system of conscription, corps, and concentration of force. In the meantime, Napoleon's economic warfare against Great Britain took him down a path that ultimately proved to be his ruin.

Six

THE EMPIRES STRIKE BACK

Symmetry Returns to the Operational Level of War

CHRONOLOGY

Fall 1806	Spain schemed against France to join her enemies.
Oct 25	French troops under Marshal Davout entered Berlin.
Nov 21	Napoleon issued his Berlin Decree initiating Continental System of Economic Warfare against Great Britain.
July 7, 1807	Treaty of Tilsit between French and Russian Empires brought general peace to continental Europe
Oct 18	General Jean Junot's army entered Spain in order to invade Portugal Napoleon deposed the Spanish king and his son at Bayonne and appointed his brother. Joseph Bonaparte became king
Nov 30	General Junot occupied Lisbon after Portuguese royal family fled.
May 2, 1808	Spanish in Madrid revolted. Murat bloodily repressed the revolt, which spread to the rest of the country three weeks later.
July 20	General Dupont surrendered to Spanish at Baylen.
August 1–21	Sir Arthur Wellesley landed in Portugal and defeated Junot at Vimeiro.

September 25	Sir John Moore took command of British army after Wellesley was recalled to Britain.
Fall 1808	Napoleon scattered Spanish forces and recaptured Madrid.
Jan 17, 1809	Napoleon left Spain and turned over pursuit of Moore to Marshal Soult.
Jan 16	Battle of Corunna; British army escaped intact from Spain.
April 1809	Austria declared war on France, Russia remained neutral.
April 19–22	Battle of Eckmühl
April 22	Wellesley arrived in Lisbon.
May 12	Napoleon captured Vienna.
	Wellesley defeated Marshal Soult at Oporto.
May 20–21	Archduke Charles defeated Napoleon at Aspern-Essling.
July 5–6	Napoleon defeated Austrians at Wagram.
July 25–26	Wellesley defeated Marshal Victor at Talavera.
October 14	Austria signed Treaty of Pressburg with France

The period after the Treaty of Tilsit (July 1807) encompassed the decline of Napoleon's Empire and his generalship—although it is often portrayed as the apogee of his Empire. It also saw the resurgence of his enemies, the empires of Austria and Great Britain (with help from the Spanish and Portuguese), who began to win tactical and occasional operational victories against him on land, and, more often than not, against his marshals. French asymmetry of arms faded as their opponents adopted the Napoleonic model of the nation-in-arms, organized along the lines of the *Grande Armée*, and mimicked French tactics and combined arms to counter the threat posed by the Napoleonic system. The period also witnessed *the defense* becoming more powerful as Napoleon's armies got bogged down fighting insurgencies in Spain, Portugal, and Southern Italy (Calabria). The French quagmire deepened when the British committed the bulk of their small army to fight in the Iberian Peninsula, using sea power for operational logistics and as an element of operational maneuver. At the same time, Austria took advantage of Napoleon's involvement in Spain to return to the contest of arms in 1809. Military historian Robert Epstein has identified this period as encompassing "the emergence of modern warfare," basing his argument on the symmetry of the 1809 campaign on the Danube between the armies of the French and Austrian empires.[1]

However, the outbreak of an extended insurgency in Spain, characterized by both regular and irregular, or guerilla, warfare, also resulted in a situation where the fundamental advantages of Napoleon's army were checked by a particularly powerful form of insurgent war that historian Tom Huber has labeled *fortified compound warfare*.[2] As manifested in Spain, this form of war saw the mixture of conventional operations by Spanish and Anglo-Portuguese armies with widespread guerilla operations over the vast, mountainous regions of Spain and Portugal. This presented their opponents with a dilemma; Napoleon's troops could either concentrate to face their enemies' armies or disperse to control the countryside, but not both. These nations were very poorly developed in infrastructure and with little arable land and armies starved marching about the arid plains and mountains in their attempts to conduct both regular and counterinsurgent operations. The Napoleonic logistical system, not always efficient even in the more developed areas of western and central Europe, broke down in Spain to a much greater degree. Command, control, and communications proved a nightmare due to the guerilla bands. Although French adversaries in Spain operated under similar constraints, they had absolute control of the waters surrounding the Iberian Peninsula courtesy of the Royal Navy and fought on and for their own soil. Also, the British established key fortified lodgments that they supplied by sea at places such as Lisbon, Gibraltar, and Cadiz from whence they sallied with military force and provided havens for guerillas and regulars alike. The French could not guard the entirety of the Iberian coast, let alone the coasts elsewhere in the Empire. They could pacify the interior of Spain to some degree, but the littoral areas were readily accessible to British ships, which provided supplies and food to forces operating throughout the Iberian Peninsula.[3] With the outbreak of war with Austria in 1809, Napoleon found himself in a two-front war, with nearly 200,000 troops committed in Spain, many of them veterans commanded by his best marshals like Soult, Ney, Victor, and even the old revolutionary marshals Jourdan and Moncey.

Another way of viewing this chapter's themes is to realize that as Jominian offensive operations become ever more indecisive (especially in Spain), the power of the defense reasserted its primacy in both Spain and Austria in 1809 under operational artists Sir Arthur Wellesley, later the Duke of Wellington, and Archduke Charles of the Hapsburgs. Wellesley, a master of defensive warfare, leveraged fortified compound warfare in Spain, which used fortified safe havens.[4] Charles, on the other hand, faced Napoleon with a reformed Austrian army in 1809 that looked in every respect like the *Grande Armée* except that it lacked Napoleon at its pinnacle, although Charles had learned enough to defeat the master at Aspern-Essling. The defeat of Napoleon in open battle for the first time, added to the preview at Eylau two years earlier, forever vanquished the idea of Napoleonic invincibility.

This chapter examines the operational generalship of two individuals—Arthur Wellesley and Archduke Charles of Austria and changes from a focus on Napoleon to include his principal adversaries. Although Napoleon matched wits with Charles, he never directly campaigned against Wellington during this period, although Napoleon did give his marshals operational guidance from afar about how they should fight him. But first we must deal with the question of how Napoleon, victorious with all of Europe at his feet, came to fight against a former ally in Spain, the British in Portugal, and a supposedly prostrate Austria, presumably humbled and cowed after the defeats at Ulm and Austerlitz.[5]

THE CONTINENTAL SYSTEM AND PEOPLE'S WAR IN THE PENINSULA

Napoleon's many decisions led to what he later regarded as one of his greatest errors, the war in the Iberian Peninsula. He remarked, after the fact, "That unfortunate war destroyed me . . . All . . . my disasters are bound up in that fatal knot."[6] His "fatal" mistakes can be traced to his animosity toward Great Britain. He never completely understood how or why his actions and policies alienated that great sea power, in part because his understanding of oceanic trade and commerce was limited to a continental outlook. At Trafalgar he had lost the capability to threaten Britain from the sea, and his disastrous experience in Egypt had taught him that straying too far from the nexus of his strength in Europe to attack British interests did not work either. He thus intensified his policy of economic foreclosure of European markets to British manufactured goods as a means to leverage Britain. This policy had actually existed during the Revolutionary Wars, but because so many of Europe's ports remained open, it did little to help the French. After the battles of Austerlitz and Trafalgar, a stalemate between the land power of France and the sea power of Britain ensued. However, when efforts for peace fell apart and Britain joined Prussia and Russia in the Fourth Coalition (1806–1807), Napoleon intensified his economic warfare against Britain. With all of Northern Europe (except Russia) under his control, he issued the so-called Berlin Decree in November 1806, closing all the ports under French control (including those in Prussia) to British trade and seizing British shipping.[7] The British responded to the Berlin Decree with their own Orders in Council that closed all British ports to shipping originating from France or transshipping through French-controlled ports. Only the ports of Portugal and Russia legally remained open to British commerce on the continent of Europe. With Tilsit the Russians joined Napoleon's system, and the "continental system," as historians have come to call it, was now complete with the exception of Portugal. Napoleon now turned his eyes toward that small maritime nation. His desire to get there was hastened by the British combined army-navy amphibious action against the neutral Danes that summer at Copenhagen, an attack that had effectively denied Napoleon the ability to seize that nation's navy. Napoleon wanted to avoid having the British seize Portugal and Lisbon in the same manner, although he was more worried about a British base than any Portuguese shipping that might be seized.[8]

To get to Portugal one must pass through Spain, but since Spain was an ally and fearful of Napoleon's vengeance for its perfidy in 1806, it gladly granted passage of a French army under the command of General Jean Junot in October 1807. In order to secure his lines of communication through Spain, Napoleon began a program of inserting French garrisons into Spanish cities between southern France and Portugal.[9] Some of these actions were retribution against Spanish perfidy because its chief minister, Manuel de Godoy, had been scheming to abandon France and join her enemies in 1806. Jena-Auerstadt undermined that plan and the Spanish were obsequious in their reaffirmation of French allegiance, but Napoleon never forgot a slight.[10]

In 1807, Spain became a cauldron of intrigue as the heir to that throne schemed to depose his royal father and eliminate Godoy. Napoleon realized the problems that a change in the regime in Madrid might entail, writing to his brother Joseph in January 1808, "trouble in Spain will only help the English . . . and waste the resources that I get from that country [Spain]." Meanwhile Junot and about 1,500 famished troops (out of an original 25,000) arrived in Lisbon on November 30 just after the Portuguese royal family had fled with the bulk of its navy and its entire treasury to the protection of the Royal Navy off the coast. This episode should have been a warning to Napoleon of the operational difficulties of rapid maneuver in the rugged terrain of the Iberian Peninsula. The British threat dominated Napoleon's thoughts as he pushed more troops into Spain, including a corps under General Dupont who was given orders to proceed deep into Spain, ostensibly to counterbalance English reinforcements that had arrived at Gibraltar.[11] Earlier, Napoleon, under the terms of his alliance with Spain, had used the large Spanish division of the Marquis de la Romana in the siege of Stralsund. He now moved it into Denmark to guard against further British actions there. This action had serendipitously denuded Spain of some of the best troops in her army at a critical time.[12]

Thus, with numerous French troops scattered throughout the Peninsula and most of Spain's best soldiers far removed to the north, Napoleon decided it was time for a change in the government of Spain. In February, through ruses and deception his troops seized a number of key fortresses along the Pyrenees and elsewhere.[13] Then, after a palace coup in March 1808 caused King Charles IV to dismiss the hated Godoy and abdicate in favor of Prince Ferdinand, Napoleon used this occasion to summon all parties to Bayonne for mediation (see map 21). Once there, he deposed and imprisoned both father and son and appointed his brother Joseph, currently reigning as the King of Naples, as Spain's new monarch. The first action was celebrated by the Spanish people, who blamed the King and Godoy for their misrule and defeats by the British, believing better times would be had with Ferdinand. However, the deposition of the popular Ferdinand rankled the masses of peasants, priests, and the proud aristocratic Spanish officer corps who still believed that God appointed kings, not French Emperors. Prior to Ferdinand's imprisonment, Napoleon had peacefully occupied Madrid with forces under Marshal Murat. The presence of ever more French troops throughout Spain, including in places like the province of Catalonia that were nowhere near Portugal, also caused resentment, especially since the ill-supplied French troops often behaved in their normal rapacious and rude style toward the locals.[14]

Scholars still debate the wisdom of Napoleon's disastrous decision to impose his brother on the proud Spanish people versus a popular and native crown prince. Some

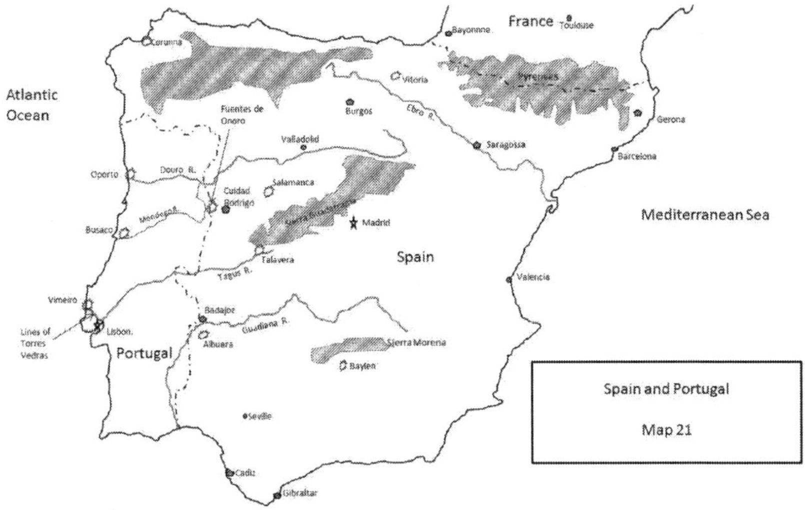

Spain and Portugal

Map 21

explain the decision as being made for strategic reasons as Napoleon moved onto his next project in his global war against Britain. He planned to dismember the Ottoman Empire with Austria and Russia and he wanted to secure the bases in Spain to move against the Maghreb in North Africa. Others highlight that it was a snap decision, made for ideological reasons. Spain was a backward monarchy ruling over a largely indolent class of aristocrats and masses of illiterate Catholic peasants. The disorder, ignorance, and lack of an ordered and efficient government in Spain offended Napoleon's sense of good government. There were some liberals in the larger towns and cities, but not many—not nearly enough as it would turn out. Resentment built and Napoleon's decision to try and enforce the Continental System at the same time as he conducted a "glorious" bloodless revolution in Spain now backfired, disastrously.[15]

On May 2, the citizens of Madrid famously rose in revolt against the hated Mameluks and other French Guard troops under Murat in Madrid. Murat brutally repressed this revolt and for about three weeks all was quiet in Spain. However, between May 20 and June 5, major revolts broke out in nearly every city and town garrisoned by the French. Aside from Junot in Portugal, which also was in a state of low-level revolt, Napoleon had over 130,000 troops scattered throughout Spain. His strategy to have Murat and his subordinates break up the organizing Spanish armies, composed of the disgruntled and disenfranchised Spanish officers of the old Royal Army, failed.

The French situation in the Peninsula went quickly from bad to worse. Conventional Spanish armies were being raised in every corner of the country and the arid countryside had now become dangerous for anything other than organized French bodies of troops as the Spanish and Portuguese peasants organized to harass and murder the hated, pagan French oppressors (as they saw it). The British government also decided to take advantage of the moment and dispatched a British expeditionary force to attempt to retake Lisbon. It was to be commanded by two elderly British generals, Sir Harry Burrard and Sir Hugh Dalrymple. Meanwhile, French arms suffered disaster in the southernmost province of Andalusia. By this time Murat, who had become seriously ill just as the revolt was breaking out, had turned over his command to one of Napoleon's aides, General Jean-Marie Savary. At the end of May, just as revolt was breaking out everywhere, General Dupont's corps had proceeded south toward Cadiz. He had gotten as far as Seville, which his men plundered, before being recalled by Savary. Loaded with loot, he dithered for much of the month of June south of the Sierra Morena Mountains as another column of French troops moved south from Madrid to join him. Dupont had little idea of how exposed his command was. Through a series of miscues and poor French decisions, rather than any outstanding Spanish generalship, two Spanish armies totaling over 40,000 men managed to place themselves between Dupont and his escape route north at the town of Baylen. The column under General Vedel that was to join him had marched north and left no guard at Baylen to keep the escape route open. After weak attempts to break through, Dupont surrendered his force of over 11,000 tired and dispirited troops on July 20 to the Spanish. The Spanish threatened to massacre Dupont and his men if he did not recall Vedel, who was escaping to the north. In a stunning display of cowardice and incompetence, Dupont ordered Vedel to surrender and that individual stupidly obeyed. The total bag in prisoners was over 17,000 French. Europe was stunned on hearing a French corps had been annihilated (literally, as it turned out, when the Spanish reneged on every promise and sent their prisoners to die of disease and neglect on prison hulks at Cadiz and elsewhere).[16]

ENTER WELLESLEY

Elsewhere in Spain, French troops found themselves cut off in cities like Barcelona or placed them under siege, as at Saragossa in Aragon. Fortunately for Napoleon Spanish efforts were ill-coordinated, but his troubles were not over. While Dupont stumbled into catastrophe in southern Spain, an English expedition sailed for Portugal. When the catastrophic news of Baylen arrived, Savary advised King Joseph, only recently arrived in his new capital, to abandon Madrid and move north to combine with the victorious forces of Marshal Bessières in Castille. On August 1, the French retreat began. That same day the British expedition arrived off Mondego Bay, about 100 miles north of Lisbon, and began landing 14,000 British regulars under the temporary command of Sir Arthur Wellesley.[17]

Wellesley came from the British nobility of Ireland and was the younger brother of the powerful and influential Richard Wellesley, Lord Mornington, scion of the Wellesley family. Wellesley had faced considerable challenges in his military career, due to the competition with other ambitious British officers who were often his seniors and the varying fortunes of his political allies, especially his brother Richard. As a younger officer he had participated in the disastrous British campaign of 1794–1795 in the Low Countries and according to one biographer learned that, "war was a serious business which should be undertaken in a thoroughly professional spirit or not at all. Having discovered what defeat was like, he gained a fierce determination to avoid it in [the] future."[18] Wellesley's apprenticeship in arms continued as he followed his brother to India and gained a solid reputation with his "sepoy" victories at Seringapatam (where he was not even in command) and Assaye. Wellesley's Indian interlude served him well for the remainder of his life. It provided him invaluable experience commanding a hodgepodge of regular British and sepoy troops, irregular native forces, in coalition with sometimes feckless native allies. It was in this environment that Wellesley first operated in both the political and military worlds. In short, it served as the ideal education for his later operations in the complex environments of Spain and Portugal. After his return to Britain in 1805, Wellesley gained further valuable experience in combined army-navy operations, serving as the second in command of the British expedition against Copenhagen in 1807 that removed that unfortunate nation's fleet from Napoleon's grasp.[19]

Wellesley shared another trait of great modern commanders, his ability to set out his ideas in writing: "Wellesley's unusual confidence in marshalling evidence and setting out an argument in long, detailed memoranda—which had been evident ever since his early days in India a decade before—made him particularly useful as a military advisor."[20] From his experiences in India and Denmark, he had demonstrated diplomacy and tact and was always careful about his logistical arrangements—key qualifications for success in operational art. Wellesley was a hard worker, and showed (like Napoleon) that the great generals in modern times must also have a talent for hard intellectual work that is not always glamorous. However, Wellesley had significant obstacles to overcome in finally being appointed as the overall commander of British forces in Spain. He was neither esteemed by the head of the British army, the Duke of York, nor, for that matter, by King George III. Wellesley had to prove himself worthy as opportunities came his way and he made the most of every one of them. Now, at the head of 14,000 troops in Portugal his time had come, but he had to hurry. There were two British officers senior to him who could supersede him at any moment, Generals Burrard and Dalyrymple.[21]

Wellesley's opponent, General Junot, now had an army of 25,000 in and around Lisbon, so the task before this impatient Englishman was of no small magnitude. His only secure route of retreat was to return to the Royal Navy ships that had disgorged his force in the first place. However, these same ships also secured his seaward flank and provided him a relatively reliable source of supplies as long as he hugged the coast in his southward advance to Lisbon. By August 5, he had begun to move south and on the August 15, collided with Junot's advance guard, which fought a stiff rearguard action at Rolica. Undeterred by this energetic French opposition, Wellesley's operational instincts led him to conclude that aggressive maneuver was his best course. He pushed his smaller army south to a strong position at Vimeiro where he learned he was about to be superseded by General Sir Harry Burrard. His impetuous advance had the unintended positive result of luring Junot out from Lisbon with an approximately equal number of troops. On August 21, Wellesley defeated his first French army, repulsing Junot's spirited attack at Vimeiro and suffering only light casualties. However, Burrard superseded Wellesley after the battle, ignored his advice and cancelled orders to aggressively pursue the beaten and disorganized French. The next day, General Dalyrymple landed and assumed command.[22]

The British had trapped Junot literally "between the devil and the deep blue sea." Their use of what is now called "operational maneuver from the sea" against an isolated French army with the right commander had yielded spectacular results. Using all his tact and flair for bluff, Junot convinced the two elderly British generals, against the advice of Wellesley, to agree to the infamous Convention of Cintra, whereby the French evacuated Lisbon and Portugal in return for being shipped back to France with "weapons, colors, and baggage." Wellesley unwisely affixed his signature to this document, probably because he believed that this was the better part of wisdom given the military talents, or lack thereof, of his two superior officers. Had he not signed it they might have continued their lackluster performance and then perhaps Lisbon would have remained in enemy hands as these two old men dithered about until another French army arrived. This fear, as it turns out, was probably justified, since Napoleon had decided that Spain needed his personal attention in order to turn the tide of defeat that had overwhelmed French arms. In any case, that September the British dutifully began moving Junot's entire force of over 25,000 troops back to France. Of these troops, 22,000 had returned to Spain by the end of the year and ended up being involved in the siege of Saragossa.[23]

Once the terms of the Convention of Cintra were revealed to the British public, all three generals were ordered home to face an official inquiry.[24] While Wellesley fought for his professional life back in London, the British army came under the command of the highly competent and professional Sir John Moore. While Moore set about organizing a new Portuguese army and ensuring that nation's security, Napoleon assumed personal command of the forces in Spain that fall. The details of Napoleon's subsequent campaign need not concern us here. Suffice it to say that the asymmetry between French arms and the hastily raised and poorly led armies of Spain again came to the fore. The Spanish exacerbated this situation by letting Napoleon reinforce and reorganize his forces in Northeastern Spain to approximately 130,000 troops. After wasting the entire month of October, they advanced loosely with four armies totaling about 120,000 troops on the area of Napoleon's concentrations in the vicinity of Pampeluna—Vitoria—Bilbao. French forces defeated them in detail, driving them in divergent directions so they could not support each other. One incident worth remarking on involved another example of British operational maneuver from the sea—gone to waste in this instance. Recall that the Marquis de la Romana commanded over 15,000 of Spain's best troops, veterans of the war in North Germany and Poland, garrisoning Denmark against further British action there. That fall, the Marquies de la Romana contacted British agents and managed to arrange to have over half his corps evacuated out from under the nose of the French by the Royal Navy. These troops returned to Spain and became the core of the army commanded by General Joachim Blake. However, many of these men were killed or captured by Marshal Victor at the Battle of Espinosa on November 10–11 as Napoleon resumed offensive operations.[25]

Scattering the central Spanish army, Napoleon advanced rapidly in the harsh winter weather to retake Madrid on December 3. Rumors of a secret mobilization by his old enemy Austria started to reach him, and he found himself now trying to manage several military and political problems at once. Meanwhile, the other French armies advanced on his flanks, one placing Saragossa under siege and the others advancing into the northwestern provinces of Spain. It was at this point that the British entered Napoleon's calculations, coming to the aid of the Spanish. Moore was soon at Salamanca and threatening Napoleon's rear and endangering Soult's corps to the north. The Spanish army in that region, now under Romana was supposed to cooperate with Moore, but together both proved too weak to do much of anything except delay Napoleon's pacification of the rest of Spain and the upcoming second invasion of Portugal. This is precisely what Moore accomplished. Napoleon raced north in an effort to annihilate the only British field army of any size anywhere in the world between his forces and those of Marshal Soult. However, the cagey Moore managed to elude the trap and retreated for the safety of Coruna at the very northwestern tip of Spain where he intended to make good his army's escape via the Royal Navy. Napoleon turned over the pursuit to Marshal Soult on January 6 and departed for France on January 17 in order to address plots in Paris and the perfidy of the Austrians, who seemed bent on war. The British retreat to Coruna in horrible weather was equally hard on both the pursuer and the pursued. British discipline broke down and when Moore arrived at Coruna, his transports were not there. By the time Soult arrived, Moore had restored some discipline to his army and occupied a strong defensive position. Although outnumbered, Soult gamely attacked and was repulsed on January 16, but Moore was mortally wounded. Nevertheless, his army was withdrawn by sea in good order. Although it had lost over 8,800 men during the brief campaign since Wellesley's departure, it gave the Spanish and British forces still in Portugal precious breathing space and prevented the outright reconquest of both Spain and Portugal.[26]

As Napoleon raced to Paris he left an active insurgency in his rear. Although the French had again prevailed over conventional Spanish armies, a key difference became more apparent as the Peninsular War (as it came to be called) went along. This had to do with the inability of the French to pacify the countryside as they continued to advance and defeat each Spanish army in turn. Often the defeated Spanish troops dispersed into the numerous mountains of Spain where they joined bands of guerillas who mercilessly preyed on French communications and stragglers. These sanctuaries were another feature of what historian Thomas Huber calls fortified compound warfare.[27] Thus it was in the Iberian Peninsula that the French had to fight not only regular conventional forces but also irregular guerrilla bands that plagued their rear areas and communications. Additionally, both guerillas and conventional forces often had sanctuaries to which they could retreat and build up combat power before returning to the fight. Often the prime example of this form of warfare are the Anglo-Portuguese forces under Wellesley, who turned the area around Lisbon into a fortified sanctuary in 1810 when Napoleon sent Marshal Masséna to retake Portugal. However, the war was really a series of localized provincial wars with sanctuaries in nearly every region of Spain. Some Spanish provinces were never taken at all, like Grenada and Murcia, and served as areas from which new Spanish armies sallied forth. When they did this, the French, who would spread out to pacify the countryside and battle the guerillas, had to concentrate and fight these armies. When this happened, the guerillas again occupied the terrain the French had vacated, punish collaborators, wipe out small French detachments, and generally interfere with French communications and supplies. As long as the British had command of the sea, they could supply these guerillas. As long as the guerillas had sanctuaries and supplies, they could fight. The larger the sanctuary, the larger the insurgent force. Thus, Wellesley's Anglo-Portuguese army was not the only conventional force fighting compound warfare in Spain, numerous conventional Spanish armies were, too, for the entire period up through 1813. If we understand this dynamic, it will help us understand why Wellesley, with armies generally no larger than 40,000 men, was rarely opposed by larger French armies, despite the French having over 250,000 troops in the Peninsula at the height of the war in 1811.[28]

Thus, keeping all these factors in mind, the Napoleonic operational way of war, as we have described it, proved unsuited to compound war in Spain. It was designed for rapid maneuver with overwhelming force and firepower, using decisive battle to scatter and demoralize its foes, and relying on short campaigns to overcome the logistical challenges of feeding the large numbers of troops required for this approach. Like the Americans 160 years later in Vietnam, it would win just about every battle but lose the war. One can compare the North Vietnamese army to the conventional British, Portuguese, and Spanish forces and the Viet Cong to the guerilla bands. French logistics collapsed in Spain, while the watery highways off its coasts ensured a logistical flow to their enemies. Poor roads, poor weather, and an agrarian base that could barely feed itself much less foraging troops undermined the French ability to maneuver rapidly at the operational level. Thus, the French found they could advance rapidly to an objective, only to run out of supplies when they got there. It was either withdraw or starve for many a French operation. Additionally, the problems with food and logistics often prevented them from massing for the same reason, as soon as two or more corps concentrated in a small area of real estate, the French found themselves unable to feed everyone. Finally, as mentioned in chapter 5, Napoleon made all the important operational decisions for the main theater of war. This worked fine when only one theater was important, but with the campaign in the Iberian Peninsula, Napoleon could not overcome the vast distances. He could beat British and Spanish armies in one province but not in all provinces. Once he left Spain he had to rely on his generals and marshals, but he still tried to control their operations, despite the great distances. Even when Europe was at peace after 1809 he still tried to control the operations of all his commanders in Spain except in the most extreme circumstances and only for generals he trusted implicitly. Only three generals ever achieved that honor: Masséna, Soult, and Suchet. Of the three, Suchet did best, but he only pacified, ultimately, three provinces. The French never solved the difficult problem of Wellesley's Anglo-Portuguese army operating conventionally in Portugal and Western Spain.

Returning to Wellesley: once cleared of the stigma associated with the Convention of Cintra, the British government turned to him as the sole commander of the effort to be made in Portugal after Moore's tragic death at Coruna. He had, after all, won the only battle of any size the British had fought to date in the Iberian Peninsula and because of Moore's daring campaign the British were still in control of Lisbon. It was not at all clear, however, that the British would even stay in Portugal. With the outbreak of war on the continent, and the chance of a decisive defeat of Napoleon by the Austrians, the topic of where to locate limited British efforts came to the fore again. Fortunately for Wellesley the decision was made to appoint him to command in Portugal with the caveat that "Lord Castlereagh [Wellesley's sponsor in the government] will keep in view that if the corps in Portugal should be further increased hereafter, the claims of senior officers cannot with justice be set aside."[29] Thus, Wellesley knew that any hint of problem or arrival of some other senior officer could mean the end of his command. He was determined not to let this happen. Meanwhile, the British government fielded an even larger military effort to aid the Austrians by landing on the island of Walcheren in Holland with 40,000 troops in July.[30] This expedition, another wasteful British effort against Napoleon's "back door," did little to tie down the French troops, but it did allow Wellesley time to cement his position in Portugal and Spain so that the idea of relieving him became ever more remote.

Earlier that March, before leaving England for Lisbon, Wellesley outlined for the King his operational scheme for the Peninsula. It is worth quoting the first paragraphs:

> I have always been of opinion that Portugal might be defended, whatever might be the result of the contest in Spain; and that in the meantime the measures adopted for the defence of Portugal would be highly useful to the Spaniards in their contest with the French. My notion was, that the Portuguese military establishments, upon the footing of 40,000 militia and 30,000 regular troops, ought to be revived; and that, in addition to these troops, His Majesty ought to employ an army in Portugal amounting to about 20,000 British troops. . . . Even if Spain should have been conquered, the French would not have been able to overrun Portugal with a smaller force than 100,000 men; and that as long as the contest should continue in Spain this force . . . would be highly useful to the Spaniards, and might eventually have decided the contest.[31]

This memorandum stated the totality of his operational schema for the next four years—a defense in Portugal could be maintained with existing British manpower and could also keep the embers burning in Spain. He especially wanted a large establishment of cavalry, artillery, and engineers, thus showing his understanding of the heavy logistical and intelligence requirements for such a campaign as well as to balance those weaknesses in the Portuguese army. He also saw the need to balance out French military prowess with extra firepower. Notice its failure to explicitly mention support to irregular forces; Wellesley had yet to factor these into his operational calculus. This memorandum was remarkably prescient; as an early form of a successful war plan it has few rivals.[32]

Wellesley departed England on April 15, 1809 and arrived to tumultuous acclaim by the citizens of Lisbon on 22nd of that same month. He was in some hurry due to Marshal Soult's invasion of northern Portugal and the capture of the second largest Portuguese city of Oporto.[33] Upon arrival, Wellesley spent a week in conversations with the Portuguese government and resident ministers and made one of the most momentous decisions of his time in the Peninsula when he appointed William Carr Beresford to oversee all aspects of the training and organization of the new Portuguese army. The Portuguese government appointed both men as marshals in the Portuguese service and Wellesley assiduously supported Marshal Beresford's authority in his training mission. This mission yielded a rich harvest in time by creating an effective and professional Portuguese army that soon became a major component in Wellesley's army and critical to its subsequent success. It serves as a model for what today is known as foreign internal defense in U.S. Army doctrine.[34]

Wellesley's next series of operations are justly famous and highlight his offensive character as an operational artist. Deciding that Soult was in fact dangerously isolated and ripe for a masterstroke, he rapidly marched his 20,000 British north from Coimbra on May 7 and arrived south of Oporto on May 12 with the wide Douro River presumably an impassable barrier between himself and Soult (Soult had destroyed the bridge upon receiving intelligence of the British offensive). Soult believed his forces safe because he had denuded much of the south bank of the river of boats and bateaux that could be used for an assault crossing. His forces could crush any lodgment on his side of the river before it could be reinforced. Thus, as with all generals who rely solely on the strength of their position for defense, he let his guard down and underestimated what an energetic and daring general might accomplish. Meanwhile, Wellesley sent Beresford with a force of mostly Portuguese regulars on a wide flanking movement to the east to unite with a small Portuguese army that the French had recently thrown out of the town of Amarante. The goal was to cut Soult's line of retreat to the east toward strong French forces in Castile and drive it north and out of supporting distance into the wilds of Galicia.[35]

Wellesley's assault across the Douro on May 12 under the very noses of the French was a model river crossing. He caught the French entirely by surprise and by the time they realized the threat he had strongly defended bridgeheads in two locations. Soult immediately saw his danger and evacuated Oporto to the northeast. On this route he seemed to be retreating exactly the way that Wellesley hoped to have Beresford block, but Beresford had not yet closed the trap. Here luck came into play—the French force protecting this line of retreat through Amarante had unwisely attacked the local Portuguese forces and been defeated and then abandoned the bridge to the enemy. Soult was forced to retreat further north. Wellesley now contemplated destroying an entire French corps and possibly capturing a famous marshal. But Soult was best when cornered; he abandoned all his baggage, wounded soldiers, and artillery and led his troops over rough mountain tracks to safety at the Galician town of Orense on May 19.[36] Soult's corps, initially 15,000 strong, had lost at least 2,000 men in these operations as well as most of its horses and all of its cannon. Wellesley noted in one of his dispatches that the roads were strewn with French stragglers murdered by the local peasants, "the natural effect of the species of warfare . . . in this country." Wellesley further estimated Soult's losses in total as at least a fourth of his army and that it could not possibly be able to resume active operations anytime soon.[37] On this score Wellesley, like so many others, turned out to be mistaken. The operational durability of a French *Corps d'Armée*, in fact, came to haunt him in his very next campaign.

With Soult presumably removed from active operations for the foreseeable future, Wellesley turned his attention to the corps of Marshal Victor threatening to invade Portugal along the line of the Tagus River from Spain. Wellesley had left some British troops at Abrantes, although his orders from London prohibited actual operations inside Spain. Not far from there, the defeated army of the Spanish general Don

Gregorio Cuesta attempted to recover after its disastrous defeat earlier that year at Medellin.[38] Wellesley obtained permission to join with Cuesta for an offensive down the Tagus toward Toledo and Madrid. It is not clear if Wellesley aimed at a spoiling operation to further prevent an invasion of Portugal that year (1809). But his purpose included developing a military relationship with his Spanish allies and a joint offensive might help their cause in many ways elsewhere in Spain.[39]

The campaign that summer proved an education for all concerned. Wellesley hoped that the dispersed nature of the French forces facing him might give Allied arms a reasonable chance for an operational success. Also, the Fifth Coalition seemed to be winning its war against Napoleon, especially given Napoleon's recent defeat on the Danube in May at the Battle of Aspern-Essling.[40] Cuesta had approximately 30,000 troops, although of varying quality in both soldiers and leaders, and another Spanish army of 20,000 faced General Sebastiani's corps in La Mancha. Wellesley did not reckon that the corps of Ney, Mortier, and especially Soult in northern Spain could provide any aid to Victor and Sebastiani who guarded King Joseph (now in Madrid again). All told, the Allies theoretically faced 50,000 French with 80,000 Anglo-Spanish troops that included 25,000 veteran British soldiers, fresh from a recent victory. But Wellesley did not count on the pride and jealousy of his Spanish allies. He also learned, to his dismay, that the Spanish promises to support him logistically were empty and he almost cancelled the campaign on these grounds alone. The elderly Cuesta did not want to take orders from a young British general (Wellesley had just turned 40) and wanted to capture Toledo and possibly Madrid, often ignoring Wellesley's advice. One moment he was all caution, the next rash. Worse, the army facing Sebastiani failed to pursue that general when he left a small rearguard and came west to reinforce Victor. King Joseph, meanwhile, ordered the three corps of Mortier, Soult and Ney (under Soult's command) to abandon Galicia and cut off the Anglo-Spanish retreat to Portugal. A potential disaster was in the making.[41]

By July 23, the Anglo-Spanish army had come as far as Talavera where it caught up to Victor. That marshal wisely retreated to the east to join forces with King Joseph's small reserve and Sebastiani's corps. By the end of the month, Wellesley had established a defensive position at Talavera as Cuesta had advanced alone against the French. Cuesta soon learned he faced an army of 46,000 veteran French soldiers under Marshal Victor who had already beaten him earlier that year at Medellin. He promptly fell back and joined Wellesley at Talavera. There Wellesley fought a two-day defensive battle on July 27–28 against Victor's army, probably the best French army that Wellesley faced in his entire career in Spain in terms of the quality of the troops. The result was a tactical victory for Wellesley. It had been a hard-fought engagement, especially on the first day when Victor had some momentary success against the British left, and the British cavalry had performed poorly. The Spanish too had been shaken, but after the first shock of battle, settled down and held firmly in their position on Wellesley's right flank. It was the bloodiest engagement for the British army in almost 100 years. The French lost approximately 7,600 men, the British 5,300, and the Spanish around 1,200. Victor reluctantly pulled his defeated forces back toward Madrid in good order.[42]

Now the combined corps of Soult, Ney, and Mortier moved against Wellesley's route of retreat. Fortunately, Cuesta had placed some troops guarding the key pass through the mountains south and these notified the Allies of the approaching danger. Wellesley withdrew his forces to safety up the Guadiana River, leaving Cuesta to his own devices in the area of Spain to the south of the Tagus. The British public and government, still celebrating Wellesley's victory at Oporto, now feted him as the hero of Talavera. It was at this time that he was elevated to a peerage, becoming Viscount Wellington of Talavera.[43] It was a hollow consolation for a failed campaign. Wellington (as we shall now call him) now retreated through some of the most barren country in Spain toward Portugal and his army suffered accordingly, especially from lack of supplies. As Wellington brought his sullen and underfed army back to Portugal, he pondered the operational lessons he had learned: that coordination with the Spanish proved difficult and that he must have absolute command over the forces under him; that he must carefully arrange for his logistics as he advanced through the famished interior of Spain; that he must secure his flanks; and that his army could not remain long in the Peninsula if it engaged in the sort of slugfests that Napoleon and his marshals were willing to fight. He innately recognized that the contest in the Peninsula depended on continued support from back home and the steady application of pressure from a position of strength, in short that operating on the strategic defensive was probably his best course of action for the future—fighting a campaign of "erosion," or attrition. He also learned of Napoleon's triumph on the Danube at the hellish Battle of Wagram, the bloodiest battle in European history since Lepanto (1571). He knew well enough that he needed all his skill and luck to face the coming storm in 1810 as new French forces would surely be diverted to Spain to cure once and for all the "Spanish Ulcer."[44]

OPERATIONAL ART ON THE DANUBE

When we last heard from the Emperor of the French, he was racing north to deal with an Austrian rebellion against his European hegemony. We must now rewind the clock and address the primary theater of war in 1809 along the Danube and consider what had changed in warfare that made this year a turning point in the history of war. The argument is made best by Napoleonic scholar Robert Epstein, who contends that the Austrian army of 1809 represented a symmetrical counterweight to Napoleon's system, opposing him with a force designed specifically to take advantage of those factors which hitherto the French had benefitted from as a result of the changes wrought by the French Revolution and Napoleon. Epstein claimed: "Napoleon would face a modern army rather than an antique one for the first time in his life."[45]

To understand the new Austrian army's ability to challenge the *Grande Armée* on relatively equal terms, we must return to the disasters of Ulm-Austerlitz and their aftermath. Recall that Austria signed a humiliating peace in 1805 with Napoleon, parting with large territories as Napoleon aggrandized his own empire as well as the domains of his allies, such as Bavaria. Nothing serves so well for meaningful military reform as failure. Such had been the case with France after the Seven Years' War and soon, as we shall see in chapter 7, with the Prussians after the catastrophe of 1806–1807. For the Austrians the interwar reform period came between 1805 and 1809. The key individual in this regard was Austria's most competent military professional, the brother of Emperor Francis, Archduke Charles.[46]

Charles had tried to implement reforms prior to the War of the Third Coalition, but had been stymied in many of his suggestions, although he and Mack had managed to at least formalize the use of the combined arms division. They also, at the last minute, had established a formal, permanent general staff.[47] However, more was needed and the "polyglot" Empire of Austria was forced to take some of the measures taken by the French in order to once again challenge the French. The first and probably least successful reform involved conscription, which proved difficult for the Austrians given the wide spectrum of nationalities. They had medieval-era militia levies that could be called out in extreme emergencies—the *Ban* of Croatia and the *Insurrectio* of Hungary—but these were essentially untrained levies and cumbersome to activate. They played no great role in the campaign of 1809, although they might have had the war continued into another season.[48] What Charles did do was increase the size of the regular army and create a limited trained militia, the Landwehr, from the German-speaking provinces in Austria. These came about 40,000 strong and contributed 15,000 troops for use along the Danube. Charles had to hand in January 1809 an impressive force of over 400,000 trained troops that could challenge Napoleon, who had numerous forces tied down in Spain and garrisoning areas of the coasts (such as Holland) and Germany (especially Prussia).[49]

More important were the organizational changes that Charles implemented. These included creating, as had Napoleon, artillery and cavalry reserves for use both tactically and operationally. In addition, he increased the artillery to more than 700 guns and would have superiority in this arm over Napoleon during the campaign. Charles concentrated the control of this numerous artillery at the higher echelons of brigade up to the army level. Charles intended to offset French *élan* and numbers with fire power. Another French innovation the Austrians adopted was the combined arms corps, which included cavalry, artillery, infantry, and engineers. Charles additionally beefed up the Quartermaster General Staff (QMGS). The QMGS included talented officers like Josef Radetzky, whom the Emperor appointed chief of staff in 1809.

Charles also oversaw the professionalization of his logistics service, making their leaders commissioned officers in the Imperial Army. In the area of doctrine, Charles adopted French skirmishing tactics, including them in the field regulations and substantially increased the number of independent light infantry (jaeger) battalions. He also created a new tactical formation known as the "mass" composed of two companies that could advance rapidly on the battlefield prior to forming line and in doing so could defend itself reasonably well against cavalry attacks. This gave the Austrian army a new level of tactical mobility on the battlefield that proved a surprise to the French in 1809.[50]

As a leader, Charles resembled more an 18th-century general in his caution and planning. His reforms, however, gave him the tools to qualify as an operational artist despite this presumed weakness.[51] As mentioned, with this new army Kaiser Francis I now felt he could challenge the French, especially given their setbacks in Spain. Accordingly, on April 9, 1809, Archduke Charles invaded Bavaria with his new operational tool. Unlike Napoleon, he still dispersed far too much of his "mass"—sending two corps to mask the army of Prince Eugene Beauharnais in Italy, one corps to contain the Polish forces in the Grand Duchy of Warsaw, two corps north of the Danube and the remaining six corps (127,000 troops) with him south of the Danube advancing on the scattered French forces under Marshal Berthier in the bend of the Danube near Eckmühl (see map 22).[52]

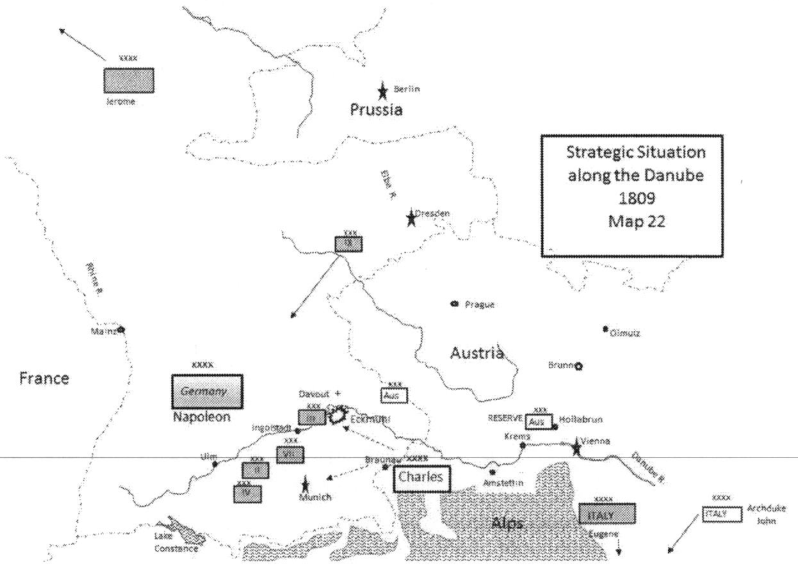

Strategic Situation along the Danube 1809 Map 22

Napoleon, still in Paris (but soon notified by his excellent semaphore telegraph system) had about 170,000 available troops—but half of them were new conscripts and many of his best generals were in Spain. Also, 50,000 of these troops were Germans, although they fought reasonably well in the upcoming campaign. Charles clearly committed too many troops to the secondary theaters, a total of over 100,000 soldiers in Poland and Italy as compared to opposing French forces who totaled fewer than 40,000. Because of this error he was outnumbered along the Danube in everything except artillery.[53] Charles's timing and location for his attack (if not his massing of his forces) proved fortuitous, especially since Napoleon had not yet joined his army. Berthier's confusing orders in the absence of his master had caused a gap in the middle of the French deployment that Charles now marched into, in some sense he had achieved what Napoleon had done four years earlier at Ulm. However, he did not realize his good fortune and instead of destroying each French corps he in turn blundered into a lengthy four-day contest against the defensive expert Davout, whose large corps fought a masterly delaying action in the vicinity of Eckmühl (April 19–22). The one positive result was the capture of the crossings over the Danube at Regensburg that allowed Charles to bring his two northern corps down into the fight as Napoleon fixed the Austrians at Eckmühl in an attempt to build up strength to move on Charles's left flank and rear. Napoleon's options were limited as his various echelons moved up to the fight. By April 23, Charles was beaten, although the battle had covered a very wide front and the Austrian artillery exacted a fearsome toll that Napoleon had not anticipated. Charles withdrew most of his forces to the north bank of the Danube, sacrificing his cavalry reserve to delay Napoleon's superior forces. The character of these operations involved corps-sized maneuver by both sides, sequential and almost constant combat over a wide front, and lack of a decisive operational victory, even though Davout, as usual, succeeded tactically.[54]

Charles retreated to Krems, his army dispirited but not destroyed. The Napoleonic paradigm of the durable operational formation, in this case Charles's new army, now applied to the Austrians as well as the French.[55] However, all was not lost. Eugene had been driven back in Italy and the entire Tyrol, which had been ceded to Bavaria in 1805, had (like Spain) risen in revolt, tying down further forces to aid in the threat to Napoleon's southern flank. Napoleon might have done well to have crossed the Danube and destroyed Charles's army, but the distractions of other theaters (to say nothing of Spain) led him to proceed as he did in 1805, along the southern bank of the Danube toward Vienna. Also, Napoleon was somewhat relieved that his army had not suffered more damage from the nasty surprise Charles had just administered. In failing to pursue *à outrance* as after Jena, the emperor gave Charles time to regain his composure and restore his army's fighting spirit and strength. Charles had advised his imperial brother to make peace after the setbacks south of the Danube had disabused him about the optimism over Napoleon's recent setbacks in Spain. Francis was made of sterner stuff and the fact that Napoleon did not pursue was perhaps evidence, which Charles had not noticed, that the French had suffered, too, from the recent combats. As if to emphasize this point, the French pursuit caught up with Charles's one corps on the south bank of the Danube at Ebelsberg on May 3 under Baron Hiller. The Austrians fought a bloody rearguard action against Masséna, then retreated across the river to join Charles, destroying the bridges as they went.[56]

The first phase of the campaign was over. The second phase now began. On May 10–12, Lannes's corps arrived at Vienna and captured those portions of the city on the southern side; however, he did not repeat his coup of 1805 and capture the bridges, thus separating Napoleon from his quarry by the wide Danube.[57] Vienna was well-stocked with provisions, which solved some of the logistical problems created by Napoleon's rapid advance, but Napoleon had not defeated the bulk of the Austrian army on the other side of the river. As Napoleon gathered his forces in Vienna for a crossing of the Danube, news came in from other quarters, both good and bad. On the plus side, Eugene had been able to force Archduke John's army away from Italy to the west and General Marmont's independent corps in Dalmatia had prevailed against the Austrian forces that had attacked it. Also, the Poles under Josef Poniatwoski had defeated the Austrians before Thorn and then invaded Galicia causing a pro-French rebellion in the province against the Austrians. More worrisome was news from north Germany, where a corps under the dispossessed Duke of Brunswick, built around troops the Austrians had allowed him to raise in Bohemia, had led a revolt inside the French puppet Kingdom of Westphalia against Napoleon's brother Jerome. Another revolt occurred in north Germany led by the Prussian cavalry officer Major von Schill. The Prussian army did not support Schill, whose revolt was put down by Danish and Dutch troops rather easily. Brunswick proved more problematic but order was restored in Westphalia with most of its new army remaining loyal to Jerome. The British, though, evacuated Brunswick's "corps" by sea, promptly transporting these forces to Spain where they came under the control of Wellington.[58]

As his forces closed up on Vienna, Napoleon, almost haphazardly, began to cross the Danube on May 20 downstream from Vienna by constructing bridges to the Island of Lobau and from there to the north shore near the villages of Aspern and Essling (see map 23). Charles moved to counter this, gathering over 95,000 troops and 292 guns just inland with a view to annihilating the French after the first corps had crossed the rivers. The Austrians had already launched several ships down the Danube to try to break the pontoon bridges spanning the river to and from Lobau. With either exquisite timing or extreme serendipity, one of these broke the main pontoon the French were crossing at midday on May 21. This event cut Napoleon, Lannes, and Masséna off on the north bank with 20,000 troops as Charles unleashed his offensive. Desperate fighting broke out all along the front at 1430 hours, just as the French repaired their bridge and Lannes's corps began to cross to join their leader.

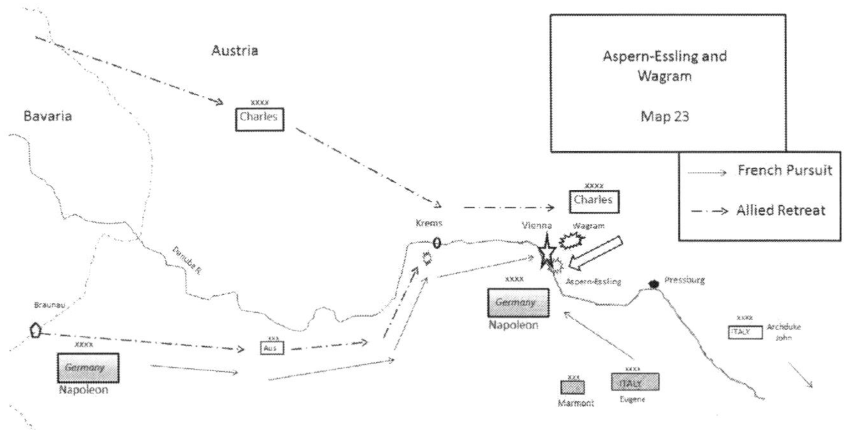

Had Charles attacked a bit earlier in the day he might have annihilated the French before reinforcements arrived, although some observers believe Charles wanted more French to cross the river before he cut them off to make the victory more complete. However, with the bridge fixed the French were able to restore the front and recapture Aspern as night fell. Napoleon was in large part saved by the strength of his position defensively, anchored by the two towns, and the fact that some of his best marshals were in the pocket with him leading the troops—Masséna, Lannes, and Bessières with the cavalry.[59]

The contest remained undecided and resumed the next day (May 22) with equal intensity. Napoleon seemed to be winning as his echelons from across the river continued to cross and reinforce his strong defensive position—it seemed another Friedland might be in the offing. However, again, Charles broke the bridge, this time with a giant floating mill. Napoleon remained trapped on the northern shore with 55,000 troops against Charles's 90,000. Ammunition began to run low and Charles personally led his men forward to attempt to complete his victory. The fighting was fierce and Lannes mortally wounded. Once the bridge was restored, Napoleon pulled back, giving Charles the first victory that anyone had ever achieved against Napoleon personally on the battlefield. The losses were immense—21,000 French and 23,000 Austrians—primarily due to artillery fire and the close-quarters fighting in the villages. The Austrian losses might have been higher if not for Charles's new tactical formation that gave his infantry a large measure of protection against Bessières's furious cavalry charges.[60]

The impact of this defeat on the two sides provides an interesting contrast between Charles and Napoleon. Charles urged his brother to make peace, believing that with this victory the Austrians could get a just settlement. Francis I did not see it that way and continued the fight, believing that he might gain more instead of losing all. Napoleon's reaction was to redouble his efforts to solve the problem of how to come to grips with Charles's army. At the same time, he had a vast empire to defend with fronts in Spain, Germany, the Balkans, and Poland all active at the same time. It is a testament to his military and political institutions that, in every case, as we saw earlier, the forces not directly under his command (other than in Spain) prevailed. Eugene defeated Archduke John at Raab on June 14. The campaign, now in its third and final phase, involved a race between Charles and Napoleon to build up combat power for the final match. Napoleon won that race in July when the army of Eugene and Marmont's corps both gave their Austrian counterparts the slip and joined him near his new and more numerous bridges on the Danube near Lobau. These detached forces constituted a second operational echelon that gave Napoleon the decisive advantage in the next round of combat.[61]

Charles's plan seemed to have been the same, to attack Napoleon's forces as they attempted to cross the river. However, with French forces in possession of a crossing further down the river at Pressburg, Napoleon employed operational deception to get the jump on Charles in crossing the river. Charles's key mistake after Aspern-Essling was his failure to assault Lobau and prevent its use by the French, since the Danube at that location was narrower and easy to bridge. On July 3–4, Napoleon began to increase the sizable forces he already had garrisoning Lobau, screening his movements with a squadron of gunboats that both protected his crossing and denied the Austrians both intelligence as well as their efforts to break the bridges to Lobau. Charles was fooled into thinking Napoleon would cross at Pressburg, but these forces, including Eugene, countermarched up the river for the crossing at Lobau. By July 3, Charles knew Napoleon would probably cross from Lobau, but even there he could not be sure where to meet Napoleon's forces given that Napoleon had occupied a number of smaller adjacent islands from which he could easily reach the north bank of the Danube. This is precisely what Napoleon did, choosing to cross the new bridges located in a different part of Lobau on July 4–5 (with a prepared floating pontoon bridge) instead of debauching on the old Aspern-Essling battlefield. Aside from covering forces, Charles decided not to contest the crossing and pulled back into a nearly 10-mile defensive position on the heights north of the broad Marchfeld plain. The subsequent Battle of Wagram on July 5–6 was a two-day affair between Napoleon's 189,000 troops and 488 guns and Charles's 136,200 supported by 446 cannon. It played out as a brutal slugfest, with artillery becoming the primary weapon of death. Never had Europe seen such an artillery barrage in its history. By the end of the second day, Napoleon had persevered through sheer, brute attrition. The Austrian cavalry reserve, decimated in the first phase of the campaign played no great role, with remounts having been hard to obtain. Overall losses for both sides were appalling,

32,000 French casualties and 40,000 Austrian, many of these from giant batteries Napoleon emplaced close to the Austrian lines to blow holes in them. Charles fell back and agreed to an armistice that allowed Napoleon a chance to recover from his bloody "victory." Francis, meanwhile toyed with the idea of continuing the fight from further east in Hungary since the Tyrol remained unpacified and the British had invaded Holland. Worse, he removed Charles from command, but had no one of equal talent to replace him with. By October, though, he had lost hope and signed a humiliating treaty (Pressburg) that stripped Austria of even more territory, made her a military ally of the French, and reduced her army to a maximum of 150,000 soldiers. Napoleon subsequently married Francis's daughter Marie-Louise in 1810 to seal the deal of the new Franco-Austrian entente and to secure an heir after divorcing Josephine.[62]

Napoleon seemed to have triumphed over Charles and the Hapsburgs. But he, too, realized something had changed. The decisive Napoleonic battle had become the attritional Napoleonic campaign. He and his adversaries learned that in this new style of warfare whoever had the most resources prevailed. Also, the temper of the battlefield had changed. Ever since the destruction of Augereau's corps at Eylau and then Bagration's corps at Friedland by massed artillery, this arm had increased in its killing power until the ultimate example of Wagram, where nearly 1,000 cannon thundered constantly for two days. One sees Sedan in 1870 and even the opening days of World War I previewed at Wagram. Napoleon made the decision to increase the firepower available to his troops after this campaign, reintroducing cannon to the lower tactical level of the regiment, thus pushing his combined arms revolution down echelon as well as up.[63] Napoleon rebuked a minister who impugned the Austrian military in later years, "It is evident that you were *not* [italics mine] at Wagram."[64] The late historian Russell Weigley and his student Robert Epstein both argue that Wagram represented a turning point, where the allure of decisive battle dimmed as battles became less and less decisive while warfare expanded in scope and became attritional. One could no longer win the campaign in one or two battles, or in battles of one day's duration. One could no longer win by destroying a division or corps, one had to destroy the entire field army, and even this might not be enough if the enemy had other armies to hand, or as a Russian operational theorist might put it, other operational echelons available in the deep rear.[65] In central Europe, as in Spain (and soon in Russia), war had assumed an absolute, *attritional* nature with these developments.

* * *

The period 1808–1809 witnessed war and politics on a scale and variety not seen in Europe since the time of Charles V or the Roman Empire. However, the canvas for Napoleon's masterpieces, to recall our metaphor, had grown too large, and his method had been copied and improved upon by the other great generals of the era. Napoleon's adversaries, who had long studied his methods, soon came into their own—excepting Wellington, who had already arrived. Napoleon, although still learning, was learning less quickly. Politically he seemed in a somewhat better position in 1810, but the war still raged in Spain and Great Britain, mistress of the seas, remained actively bellicose. Russia, which had remained guardedly neutral in 1809, soon openly rebelled against Napoleon's "antieconomic" Continental System. Napoleon's ruinous strategy against Britain was soon to make the specter of a two-front war permanent.[66]

SEVEN

NAPOLEONIC TOTAL WAR IN SPAIN AND RUSSIA

1810–1812

CHRONOLOGY

Nov 19, 1809	Marshal Soult destroyed Spain's largest organized army at Ocana.
Jan–Feb 1810	Soult conquered Spanish province of Andalusia.
Apr 17	Napoleon appointed Marshal Masséna as commander of the Army of Portugal.
Jun 10	Masséna captured Spanish fortress of Ciudad Rodrigo
Aug 27	Masséna captured Portuguese fortress of Almeida
Sep 27	Wellington defeated Masséna at Bussaco and fell back into Lines of Torres Vedras
Dec	Russia accepted British goods into its ports and secretly prepared for war with France.
Mar 1811	Masséna's army, famished for lack of supplies, retreated from Portugal.
Mar 10	Soult captured Spanish fortress of Badajoz.
May 3–5	Wellington defeated Masséna at Fuentes de Oñoro.
May 11	Wellington captured Almeida.
May 16	Beresford defeated Soult at Albuera near Badajoz.
Jan 19, 1812	Wellington captured Ciudad Rodrigo.
Apr 6–7	Wellington captured Badajoz

Jun 23–25	Napoleon crossed the Niemen River and invaded Russia. Size of French main army group became approximately 380,000.
Jul 22	Battle of Salamanca, Wellington defeated Marmont and Soult withdrew from southern Spain.
Jul 28	Napoleon was at Vitebsk. Size of French main army group became approximately 200,000.
Aug 12	Wellington liberated Madrid.
Aug 15–19	Battles near Smolensk, Russian armies united to defend Moscow. Size of French main army group became approximately 150,000.
Sep 7	Battle of Borodino, Russians under Kutusov withdrew and abandoned Moscow.
Sep 14	French entered Moscow. Size of French main army group became 95,000.
Oct 18	Napoleon's *Grande Armée* began retreat from Moscow.
Nov 25–29	Napoleon crossed the Beresina. Size of French main army group (including stragglers) became approximately 40,000.
Dec 15	Size of the remnant of French main army group, including stragglers and reinforcements became approximately 12,500.

Nations in arms, army groups, combined arms organization, tactics at all echelons, and massive fire power—by the end of 1809 all these things were in place on "both sides of the aisle" during the Napoleonic Wars. However, this study contends that, despite 17 years of nonstop conflict, Napoleonic warfare only became "total war" in 1810. Certainly, total war in the sense we use it today includes all the factors mentioned in the first sentence.[1] Allied strategists and generals unknowingly applied Hans Delbrück's strategies of attrition, using protracted conflict to wear their enemies down—in this case Napoleon and his French-led armies. Wellington had already outlined such a strategy *before* returning to Spain. In 1810, with a massive invasion of Portugal looming, he put into place the operational framework for protracted attritional warfare, policies, and actions, which may be termed *absolute* or *total*, by his troops in concert with irregular forces. Consistent with Delbrück, and reflecting the wisdom of Clausewitz that weaker powers must take advantage of the defense as the stronger form of war, Wellington showed the way in his 1810–1811 campaign in Portugal by using combat power, not for the offensive, but for the strategic and operational defensive as a means to sap the strength of an army of potentially 100,000 men invading Portugal.

Similarly, when Napoleon made his disastrous decision to invade Russia and solve the problem to his east once and for all, the Russians, less deliberately, stumbled into the power of the strategic defense as a means of exhausting the enemy. In both of these hellish campaigns—Portugal and Russia—geography played an immense role in the operational framework. But they shared these commonalties of defense, protracted resistance and retreat with a goal of "erosion," as Wellington named it, of the adversaries' larger forces. In both campaigns, the key for the operational artists involved—Wellington, Marshal Kutusov, and the less-known Tsarist generals Barclay de Tolly and Peter Bagration—centered on the Clausewitzian concept of the "culminating point of victory" that "turning point at which attack becomes defense." In *On War*, Clausewitz went further, undoubtedly using his personal observations of the 1812 Russian campaign to declare: "It is the defense itself that weakens the attack. Far from this being idle sophistry, we consider it to be the greatest disadvantage of the attack that one is eventually left in a most awkward defensive position."[2] Using operational art, Wellington and the Russians managed to engineer precisely this "awkward defensive position" in front of the lines of Torres Vedras in Portugal in 1810–1811 and Russia at Moscow in 1812.

Although they were not total war, in the modern sense that sees the phenomenon inherently related to technology, these campaigns emphasize the bedrock of human, political, strategic, and geographical components of total war. They constitute a reflection of the concept as all-out war, on the largest possible scale, and to the bitter end of exhaustion and defeat, for one side or the other. Such was World War I, whose 100th anniversary is now upon us. So also were these campaigns over 200 years ago.

* * *

Napoleon seemingly triumphed in 1809—both in Spain and against Austria. Yet, Great Britain remained at war while the Napoleonic system in Europe and a cruel compound war in Spain and Portugal simmered on. The Napoleonic system still had some triumphs left to it in Spain, but not against Lord Wellington. Wellington grimly committed himself to a strategic defensive, ignoring the pleas of his Spanish allies to join them in yet another round of fruitless offensives against the French forces occupying the key central position in Spain. His experiences in the unsatisfactory Talavera campaign and the defeat of the Austrians at Wagram convinced him of this necessity and he advised his Allies to remain on the defensive and husband their strength. He withdrew to Abrantes in Portugal to prepare for an invasion army that might have Napoleon at its head. The British polity also influenced Wellington's decision for the defense, as opposition politicians proposed in 1809 and again in 1810 to withdraw Britain's one field army from the Peninsula given the threat posed to it and the apparent French victory in Spain. By not risking his army in any further offensive adventures, Wellington managed to maintain its presence in Portugal for the time being.[3]

The Spanish pressed on without him and pushed an army of 50,000 men north from Andalusia in an attempt to retake Madrid. This army initially advanced rapidly, catching the French dispersed, but then balked along the Tagus River. Marshal Soult, in overall command, concentrated his forces and destroyed the Spanish army at Ocana on November 19. Napoleon, who had planned to come south of the Pyrenees, now became distracted by his divorce from Empress Josephine and subsequent marriage to Austrian Princess Marie Louise. These distractions provide further evidence of the decline in his political and military judgment. He left Spain to his subordinates as he pursued his dream of having a legitimate heir to his throne. King Joseph and Marshal Soult decided to take advantage of their victory at Ocana and press ahead during the winter of 1810 by conquering Andalusia, scene of the disaster at Baylen two years earlier (see map 21). In January–March, Soult moved south with over 70,000 troops in three corps. With the only effective Spanish army destroyed, he had little trouble taking Seville, causing the ruling Spanish Junta to flee to Cadiz. Soult followed and put this place under siege, but he, Joseph, and Napoleon—oblivious to the strategic mistakes being made in Spain—had now tied down 70,000 more troops to occupying a vast area of Spain without any positive result. The guerilla resistance continued. As a consolation of sorts, Soult captured the key fortress of Badajoz that dominated the southern invasion route into Portugal on March 10, 1811.[4]

By April, Napoleon had a new empress and had decided to leave the most important job in Spain, dealing with Wellington's dangerous Anglo-Portuguese forces, to his most illustrious marshal. The emperor had recently named Andre Masséna Prince of Essling for his superb combat leadership at Aspern-Essling, leadership that probably saved the French army from disaster. Masséna was certainly the right man for the job in Spain; if anyone could defeat Wellington it was him. However, Soult was now tied down occupying and garrisoning vast areas of Estremadura (Badajoz) and Andalusia, and would be no help in the upcoming campaign. Nonetheless, Masséna had over 65,000 troops in three corps commanded by Ney, Junot, and the Swiss mercenary general Jean-Louise-Ebenezer Reynier. Masséna assumed command of his force in May at Vallodolid before moving up in June to oversee the reduction of the fortress of Ciudad Rodrigo. The siege of this place, along with the Portuguese fortress of Almeida, gobbled up much of the summer. Masséna's forces had already been campaigning for two months before they pushed on toward Lisbon in earnest in early September. Napoleon's attitude toward the challenge that Masséna faced was completely unrealistic, observing "It was absurd his armies should be held in check by 25,000 or 30,000 British troops."[5]

Napoleon's assessment of Masséna's challenge must be examined against Wellington's intricate plans for the defense of Portugal and an anticipated invasion by as many as 100,000 French troops. Thanks to Soult and King Joseph he would deal with considerably fewer than that. Also, thanks to these other Spanish adventures and Napoleon's odd attention deficit when it came to the Anglo-British forces on his flank, Wellington had been given almost a year to prepare for the invasion.

Wellington's plan had four major components: creation of a suitable Portuguese regular army to augment the British; construction of a line of defenses from the broad Tagus to the Atlantic; training and equipping of the ancient Portuguese militia—the Ordenanza—for use against French communications to Spain; and, finally, the complete devastation and depopulation of the Portuguese countryside. This last served as a means to deny the French any ability to use their logistical practice of foraging their troops on the countryside and sequestering village stores.[6]

The first component, as we have already seen, was given over to Marshal Beresford, who by the time of Masséna's invasion had trained over 30,000 Portuguese regulars, some with him in the south observing the southern invasion route, but most in the north integrated into Wellington's field army of 50,000 facing Masséna. Wellington had already noted the defensive value of the terrain north of Lisbon, centering on the town of Torres Vedras. In October 1809, nine months before Masséna entered Portugal, the British commander had begun fortifying this ground, 30 miles in length from the ocean to the impassible (due to the Royal Navy) estuary of the Tagus River on a roughly northwest-southeast line. He judged that he could move much of the Portuguese population into this strategic redoubt, which reminds one of the defenses seen later on the Western Front in World War I. The line reflected a defense in-depth of three "belts" of fortifications with dozens of forts and many more redoubts, all covering various fields of fire with lateral roads along which Wellington could rush his mobile reserves to shore up the front. Eventually, these lines contained over 447 pieces of artillery.[7]

The last two components of Wellington's defensive scheme were equally important, but aided by both the geography of Portugal and Masséna's choices. First, Masséna and Napoleon had not made arrangements to secure the line of communication back into Spain, believing the countryside empty when in fact the Ordenanza could and did interdict these lines effectively. Any French formation smaller than a battalion or a squadron found it difficult to support itself against the well-armed, mobile irregular forces that Wellington and Beresford put into the field. Finally, Wellington's chore in devastating the Portuguese countryside was made easy by the arid and unforgiving region through which Masséna invaded. Instead of advancing on the south bank of the Mondego River, where the roads were better and closer to more fertile regions, Masséna invaded along the north bank of the river valley, some of the most forbidding and inhospitable terrain in the country. Masséna described Wellington's "scorched-earth" policy in a report to Berthier on his advance, "Sir, all our marches are across a desert. Not a soul to be seen anywhere; everything is abandoned."[8]

Masséna had sidestepped Wellington's plans for a prepared defense on the main road from Almeida to the interior of the country, but the Englishman reacted quickly, moving his forces opposite the French and conducting a delaying retreat until reaching a strong position at a ridge near Busaco (see map 21). Here, on

September 27, Wellington fought another of his signature defensive battles—52,000 Anglo-Portuguese defeating 60,000 French and inflicting 5,000 losses on them for 1,500 Anglo-Portuguese lost.[9] The steadiness of the new Portuguese troops proved a rude shock to Masséna; however, despite this defeat he outflanked Wellington's position. Wellington, instead of risking another tactical action, simply retreated toward his elaborate defenses at Torres Vedras. On October 10, he had over 77,000 men manning these lines, including 8,000 Spanish regular troops from the Army of Estremadura and 11,000 armed and trained militia. Wellington reinforced the lines with about 20,000 men, keeping his main field army concentrated as an operational reserve behind the lines to react to any major attack that Masséna might attempt.[10]

Masséna followed Wellington, again through an almost entirely deserted landscape with little food. Then, in a scene that foretold similar disasters to come in Russia, Junot's ill-disciplined corps arrived at Coimbra, the only major city until Lisbon on his route of march, and sacked it, reputedly destroying enough food and stores to have revictualed the entire French army and sustained it for weeks. Instead, the remainder of the French army continued on meager rations. Masséna did not yet know of the Lines of Torres Vedras and made the decision to try to catch Wellington's army in its retreat to Lisbon and force a battle on better terms than at Busaco. Perhaps he could even hustle the British aboard their ships for a hasty evacuation of all of Lisbon as Soult had done at Coruna. Accordingly, he left 3,500 of his wounded and sick in Coimbra with only two companies to guard them. On October 7, a Portuguese militia division swept down and recaptured Coimbra with all of the wounded and sick as well as French supplies being gathered there. Shortly after Masséna learned his communications through Coimbra were cut, he discovered the nasty surprise of Wellington's fortified defensive belt. His cavalry arrived before these positions on October 11 amidst the cold autumn rains and reported the fortifications. Masséna himself reached them on October 14. The old Marshal's practiced eye informed him of the absolute strength of the position, all along its front, with the victors of Busaco waiting to administer another rebuff to his tired, hungry, and now cold troops. With the Royal Navy protecting its flanks, there was no way to outflank this position.[11]

Masséna had reached the culminating point of victory before Busaco—at Torres Vedras he was clearly beyond it. His troops were cut off, they could not go forward; but Masséna, whose strength of will had been severely tested in the past, was not inclined to retreat. The French had little in the way of supplies and were slowly starving while the Royal Navy supplied their adversaries by sea. In no other campaign did the British advantage in sea power and the French approach to logistics play as large an operational role. Masséna decided to encamp in front of the lines, hoping the British might attack him and give him a cheap victory. Wellington made no such error, relying on irregular forces, the strength of his works, and starvation to win him the victory. Now, with every day that passed, Masséna grew weaker. This lasted for almost a month before Masséna, to avoid the complete starvation of his army, pulled back up the Tagus to Santarem, where his army could forage and await reinforcements from Napoleon. It is a testimony to Masséna's will that he made it through most of the winter encamped in Portugal, but his army was ruined. There was some hope that Soult, who had begun besieging Badajoz in January 1811, might advance on the southern front and link up with Masséna south of the Tagus. But Soult's siege dragged on and by March 5 the Prince of Essling began his retreat back the way he had come toward the supplies and fortresses of Almeida and Ciudad Rodrigo. Five days later (March 10), Soult captured Badajoz.[12]

Foreshadowing his later performance in Russia, Marshal Ney commanded Masséna's rearguard during the retreat, allowing the remnants of the once mighty army of Portugal to withdraw with some sort of order into Spain. Wellington followed close on his heels but failed to achieve any major victories given Ney's tenacity and skill. Wellington was able to place Almeida under siege in April. The cost to French arms had been over 30,000 troops with the result that the British position was immeasurably stronger than it had been the year before. The French situation got no better. Napoleon's relationship with Tsar Alexander had deteriorated to a point that he had already begun to withdraw his best soldiers from Spain, which at its peak contained 370,000 troops of the French and their allies. Wellington's campaign had broken the back of the ability of the French to take the strategic offensive against the Anglo-Portuguese armies along the rugged Portuguese-Spanish frontier, as became clear in May 1811. That month saw both Masséna and Soult attempt limited counteroffensives to relieve sieges of Badajoz (by Beresford) and Almeida (by Wellington).[13]

Wellington again underestimated the operational durability of a French army. Masséna had restored some semblance of discipline after feeding and resting his tired troops (and gaining some replacements) at Ciudad Rodrigo. Early in May he advanced against the outnumbered Anglo-Portuguese army at Fuentes de Onoro with 48,000 men to Wellington's 36,000, achieving both tactical and operational surprise. Although Masséna managed to flank his enemy's position, Wellington again won on the defensive by "refusing" his right flank with some very skillful maneuvering. The victory was somewhat diminished when the garrison of Almeida escaped back to French lines on May 11. Despite this, Fuentes was Masséna's swan song; he lost some 2,800 casualties to about 1,800 for Wellington. As his master was to learn in Russia, the old fox no longer had the youthful vigor for harsh campaigning that the war in the Peninsula required. Napoleon replaced him with the younger Marshal Auguste Marmont.[14]

In the south the Beresford attempted to recapture the critical fortress of Badajoz, which was absolutely necessary before even limited offensive operations could be conducted inside Spain. Here the result was less fortuitous. Soult marched rapidly to relieve Badajoz and almost defeated Beresford's Anglo-Spanish-Portuguese army by again employing a flanking attack. The Spaniards, ably commanded by General Javier Castanos, held their own and a timely counterattack by the British Fusilier Brigade retrieved the situation, but losses on both sides were extremely heavy—8,000 (approximately 33%) for the French and 6,000 (approximately 18%) for the Allies, especially among the British (nearly 50%)—one of the heaviest loss rates of the Napoleonic wars. Badajoz, however, was resupplied and the British siege collapsed. Wellington never let Beresford command independently in combat again after this Pyrrhic victory.[15]

With these two battles, Wellington's war in Spain gave way to a stalemate that centered on the two French-held fortresses of Badajoz and Ciudad Rodrigo. Whenever Wellington tried to invest these places, the French—Marmont and Soult—combined and confronted his forces, usually using the great bridge across the Tagus at Almaraz for lateral movement. Wellington resorted to a lengthy cat and mouse game for the rest of 1811, hoping for a French error. In the meantime, Napoleon continued to pull troops from Spain, weakening his forces further. By early 1812, as Napoleon was finalizing his invasion of Russia, Wellington struck. He first feinted toward Badajoz and then headed north and besieged Ciudad Rodrigo, assaulting and capturing it on 19 January. Then, moving swiftly against Badajoz, he placed it under siege for the third time since its capture by the French, famously assaulting it on the night of April 6–7, although his victory was sullied when his troops sacked the Spanish fortress in an orgy of looting, rape, and murder of the hapless Spanish inhabitants. These two coups put the lie to claims that Wellington was not capable of decisive, even rash, operational maneuvers.[16]

With his flanks thus secured, Wellington moved to ensure that Soult and Marmont could not come to each other's aid once he advanced against one or the other in Spain. To accomplish this he sent his most trusted subordinate, General Rowland Hill, against the great bridge at Almaraz across the Tagus that connected Soult's area of operations with that of Marmont. Hill put the garrison to flight and destroyed the bridge on May 19. Wellington now commenced offensive operations deep into Spain for the first time since 1809—just prior to Napoleon's invasion of Russia. As usual he employed operational deception, tricking Soult into believing he intended to liberate Andalusia. Wellington reasoned that if he could defeat Marmont in central Spain, Andalusia must fall of its own accord. He advanced on interior lines to the north and attempted to lure Marmont's army into a battle around Salamanca. However, Marmont proved almost as cautious as Wellington. The two spent most of the summer maneuvering against each other's lines of communications in something of a throwback to 18th-century warfare (neither organized his army in corps, but rather left them in divisions for the duration of the campaign). Finally, on July 22 near the start point of the campaign at Salamanca, Marmont allowed the lead division of his army to become separated from his main body and Wellington pounced, destroying that division and scattered the next one behind it. He then repelled a vigorous French counterattack. Marmont was badly wounded and his army thrown into a disorganized retreat after losing almost 14,000 killed, wounded, and prisoners.[17]

This victory changed the landscape of the war in Spain. Soult was forced to evacuate southern Spain when Wellington liberated Madrid on August 12. Wellington overreached himself by attempting a lightning advance to seize Burgos in October. Unable to capture the city's citadel and threatened by the superior French forces under Soult moving on his rear, Wellington conducted a strategic retreat back to his starting point at Ciudad Rodrigo. The French reentered Madrid, but they had traded the entirety of Southern Spain to regain it. By now December 1812 had arrived and Napoleon's disaster in Russia made manifest, although its full import was not completely understood in Spain.[18]

From 1809 to 1812, Wellington had employed attrition warfare against the French, at one time dealing with the entire might of Napoleon's empire. French errors notwithstanding, Wellington's elegant mix of fortified compound warfare and protracted operations in the Peninsula exhausted French resources. French reinforcements seemed to arrive as worn and depleted as the troops they were supposed to support. This same phenomenon was now about to occur on a far larger scale in Russia. In contrast, Wellington's creation of a viable Portuguese army gave him military echelons that the French ignored or underestimated as at Busaco. No longer was Spain an object to be conquered; it had become the base for an offensive threat to be defended against—and at the worst possible moment for the Emperor of the French.

NAPOLEONIC TOTAL WAR IN RUSSIA

The causes for the breakdown of the seemingly amicable relationship forged between Russia and the Napoleonic Empire at Tilsit are beyond the scope of this study. However, certainly Napoleon's ruinous Continental System (see chapters 5 and 6) had done much to prepare the ground. By 1810, Russia was already preparing for war with Napoleon after Napoleon had offended the Tsar by occupying the Duchy of Oldenburg, which belonged to his sister's husband. Napoleon's policies vis-à-vis the Duchy of Warsaw and the potential political resurrection of Poland as a strong French ally further exacerbated relations between the two powers. Alexander even contemplated preemptively attacking the Duchy of Warsaw in the hope that the Prussians and Austrians might throw off the Napoleonic yoke and join him. Fortunately for Russia this did not happen. Instead, the Tsar's emissaries signed the less than favorable Treaty of Bucharest in the long-running war with the Turks, thus freeing up distant armies for use against Napoleon prior to the French invasion in the summer of 1812.[19]

As the Napoleonic-Russian imperial relationship went from alliance, to neutrality, to hostility, Wellington's campaign and use of the Lines of Torres Vedras gave thoughtful observers a recipe for the antidote to Napoleonic offensive power. Tsar Alexander watched these events unfold and commented to Napoleon's ambassador Armand de Caulaincourt in 1811:

> If the Emperor Napoleon makes war on me . . . it is possible, even probable, that we shall be defeated, assuming that we fight. But that will not mean that he can dictate a peace. The Spaniards have often been defeated; and they are not beaten, nor have they submitted. But they are not so far away from Paris as we are, and have neither our climate nor our resources to help them. We shall take no risks. We have plenty of room; and our standing army is well organized.[20]

If one equates the frozen forests and steppes of Russia to the British-dominated seas surrounding the Iberian Peninsula, the analogy is even closer to the example provided by Wellington and the Spanish. The solution to the Napoleonic method of the one-season, decisive campaign offered itself to both the Tsar and his minister of war General Michael Baron Barclay de Tolly—scorched-earth, mobilization on the largest possible scale, guerilla warfare, fighting retreats, inconclusive bloody battles, and no surrender. However, all of the elements of this strategy were not immediately adopted, even though the top leaders understood them as potential, even probable, courses of action.[21]

Throughout all of 1811 and the first part of 1812, the two sides built up their forces and prepared diplomatically for the expected "clash of the titans." Napoleon's preparations were on a gargantuan scale and by the spring of 1812 he had deployed over 449,000 troops in cantonments along the Vistula River in Poland.[22] His alliance included the military forces of just about every dominion under his direct and indirect influence, including sizable contingents from both Austria and Prussia. Neither of these contingents can be categorized as "willing" and it is to Napoleon's great discredit as a strategist and general that he had the bad judgment to place them on his northern and southern flanks to guard the advance of his central army group. The final break came effectively in April, when the Tsar presented Napoleon with an ultimatum to evacuate Prussia and return the Duchy of Oldenburg. This was tantamount to a declaration of war since it meant that Napoleon must also abandon the Grand Duchy of Warsaw. Adding injury to insult, the Tsar signed a treaty with Sweden, which freed further troops for use against Napoleon. Once Napoleon invaded, the Tsar's representatives in Great Britain also signed a treaty of alliance with that nation, although British goods were already flowing into Russian ports. However, Russia would not lack for funds to help pay for the war.[23]

The plans and deployments of the French and Russians for this campaign reflect best, perhaps, the style of mobilization and operational components envisioned by the Soviet operational theorists like Isserson and Tukhachevsky 120 years later in the

Soviet Union.[24] Russia's vast size seems to bring this sort of gigantic military effort out, no matter the age and the technological sophistication. We will turn first to Napoleon's plans and deployments prior to the invasion. Napoleon cannot be faulted for failing to mobilize his vast resources. He even tried to broker a peace in Spain with Great Britain in order to simplify the task at hand, although this too-little, too-late initiative failed.[25] Napoleon's arrangements envisioned three strategic echelons in order to maintain the momentum of his offensive should he be drawn deeply into Russia, despite his hope that he could fix and defeat the Russian armies in Russian Poland and Lithuania. Napoleon's first strategic echelon consisted of the bulk of the reconstituted *Grande Armée*, with the enlarged Imperial Guard (nearly 50,000) under his personal command. In addition were two armies commanded by his relatives, one under his step-son Prince Eugene Beauharnais, the Viceroy of Italy, and the other under his brother Jerome, King of Westphalia. These three armies constituted the central army group, also under Napoleon's operational command, and might be considered the first army group of the modern era. This army group numbered 383,000 troops and 1,022 artillery pieces. Nearly half of this force was composed of non-native French troops: Dutch, Swiss, Poles, Italians, Spanish, Portuguese, Croatians, and Germans from the Confederation of the Rhine. As mentioned, on Napoleon's north and south flanks, respectively, were Marshal MacDonald's X Corps (32,000 Prussians, Germans, and Poles with 84 guns) and Prince Karl zu Schwarzenberg's Austrian corps (34,000 men with 60 guns).[26]

The second strategic echelon consisted of troops still forming up to be used as replacements for those in the first echelon. They consisted of about 165,000 soldiers, many of them in Marshal Victor's IX Corps and included many Polish and German troops. The third strategic echelon amounted to about 60,000 troops and included the corps of Marshal Augereau, two of whose divisions were in the second echelon with Victor, as well as various garrisons in Poland. We might consider that a fourth strategic echelon existed in the remainder of the empire encompassing the National Guard, fortress garrisons and posts along the Atlantic coast (not including the Peninsula), the Danish Army and a virtual levée en masse for any French males not already serving between the ages of 20 and 26. These constituted something less than 275,000 men, but Napoleon was relying on further exactions from his allies—especially the Germans—and new troops to be raised in "liberated" areas of Russian Poland and Lithuania should he need troops for an unexpected crisis. Napoleon's mobilization of these numbers could not be maintained for a protracted war, feeding and billeting them soon proved so daunting that they must either advance or demobilize—by May 1812 Napoleon had decided they would advance.[27]

Napoleon's plan was characteristically simple and based on reasonably good intelligence about the size and disposition of the two main Russian armies (discussed in the following). His goal was nothing less than the destruction of the bulk of Russian military power in Lithuania or at the very least west of Smolensk using his main army group. The larger Russian army (under Barclay de Tolly) numbered, according to his intelligence, around 125,000 troops as compared to a smaller army of 50,000 troops to its south (under Prince Bagration). The Emperor's plan was to advance directly on the larger army, fix it, and destroy it. Despite having identified the location of the larger Russian field army prior to hostilities, successful execution of his plan required good intelligence about its position and movements once the campaign started. Napoleon simply decided to aim at the center of Barclay's concentration of forces by marching on Vilna, the capital of Lithuania. A second assumption involved his being able to outmarch the main Russian army with at least enough of his troops to get it to stop and fight. Napoleon intended to improvise operationally as the situation dictated, hoping to conduct a massive maneuver against the flank and rear of Barclay's army with either reserves from the main army or those under Eugene, the army on his immediate right. Finally, since the other two armies were positioned in echelon on his right flank, the Emperor planned for them to at least keep the more southerly Russian field force from joining with the main northern one. Once the destruction of the main Russian army was accomplished, Napoleon could turn to destroy the southern army, especially if Eugene or Jerome managed to find and fix it or slow it down, and then send out peace feelers to the Tsar. If these were rejected he could then bring up the second echelon and advance toward either St. Petersburg or Moscow. If all went well the campaign would be over in six weeks, somewhere in the vicinity of Vilna in Lithuania. If not, he could continue the campaign as necessary to a successful conclusion by the end of the summer or in early autumn before the first snows. Surely the Tsar would surrender if his armies were destroyed and one of his two capitals had been occupied? The willingness of the Tsar to surrender constituted Napoleon's final assumption even though Caulaincourt had reported differently.[28]

Napoleon's logistical preparations to support this plan made sense had the campaign turned out as he planned. Although his troops would be required to forage in some cases, he had large amounts of biscuit, grain, and mobile ovens created to support his armies. Also, his line of advance, and indeed his initial penetration into Russia at Kovno at the juncture of the Nieman and Vilia Rivers, was intended to take advantage of those two streams to support the main line of his advance to Vilna (see map 24). The weak link in the entire system of logistics, especially if the campaign did not develop along the nice, neat lines he hoped for, involved distributing the food to the immense army from river debarkation points and supply convoys. The means to do this involved using draught animals and these animals in turn needed as to be maintained and fed. Thus, a key problem might arise if the Russian army managed to escape to the east of Vilna without an engagement. Napoleon would be forced to choose between two options. The first required spending the precious weeks of good campaigning weather establishing a new line of logistical support using rivers and the

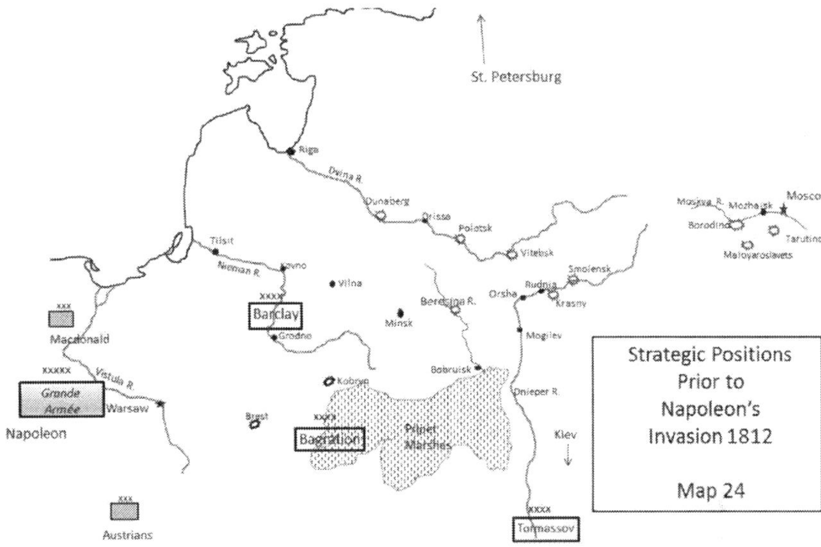

Strategic Positions Prior to Napoleon's Invasion 1812

Map 24

Baltic toward St. Petersburg or, more likely, along the Dvina and Dnieper (Dnepr) rivers toward Smolensk. The second option would be to outstrip his logistical support by rapid marching of his troops on either line of Russian retreat in the hopes of catching them, while foraging as a stopgap measure. This last choice was risky, and Napoleon knew of it based on his previous experience in 1806–1807. Finally, the assumption of good weather during the Russian summer did not account for an extremely dry, hot continental summer. As we shall see, the extreme summer weather was to do more damage to Napoleon than either of the two Russian armies until the Emperor arrived before Borodino in August. Also, these circumstances applied somewhat to the Russian armies with the exception that they could forage far more effectively in their own territory than Napoleon and would be retreating onto their lines of magazines and depots, which permitted them to destroy what they did not eat or carry. Things would be immeasurably worse for Napoleon if the Russians employed a scorched-earth campaign along the route of their retreat as Wellington had in Portugal.[29]

The Russians had a strong hand to play despite the nearly half million-man field army Napoleon had deployed for his campaign. They initially matched it with over 200,000 men in their first strategic echelon. As we have seen, they had two main field armies under Barclay de Tolly (hereafter Barclay: 127,000 men, 558 guns) and Prince Peter Bagration (48,000 men, 217 guns). There was a third army under General A. P. Tormassov (42,000 men, 164 guns) stationed south of the Pripet Marshes to protect Kiev and the southern invasion route of Russia.[30] Alexander, based on his experiences in 1805–1807, had appointed Barclay minister of war in 1810 and adopted his scheme for mobilization that verged on the French idea of the levée en masse. These measures resulted in almost 800,000 men having been mobilized prior to Napoleon's invasion, but, as seen, only 200,000 of these were trained and immediately to hand in June 1812. The rest were in the additional echelons available over the long term to the Tsar or stationed along Russia's vast imperial borders. First, the Tsar's diplomacy had freed up substantial forces from the Turkish and Swedish fronts: the Finnish Corps of General Steinheil (18,000 men, 72 guns) and the Army of the Danube under Admiral Tshitshagov (53,000 men, 240 guns). When added to other second echelon formations the total came to nearly 137,000 men, but most of these would not come into play for operational use until the late summer. Garrisons, depot battalions, and squadrons, Cossacks not yet called up, and some militia forces constituted a third strategic echelon of about 161,000 men, of which 133,000 eventually came into action against the French. Napoleon's plan of mobilization theoretically could match (and overmatch) the arrival of these additional troops, but it is worth keeping in mind that almost 70,000 of these troops were veterans from Finland and the Danube and not second-class troops in any sense. Worse, Napoleon only learned of their availability after he had crossed the Nieman on June 24–25, 1812.[31]

From the beginning the Russians favored a Fabian strategy, avoiding direct engagement with Napoleon's main army.[32] To understand how the final plan came about, one must first understand the Imperial Russian army of that era. The shadow cast by Suvorov proved long and self-appointed candidates to succeed him were many. Also, Russian army leaders had played a key role in the assassination of the Tsar's father Paul I, and this explains why the Tsar was unable to support Barclay as strongly as he might have once the campaign began. Barclay developed his plan principally from ideas that a Prussian named Ludwig Wolzogen advocated in 1810—demonstrating how early Russia had begun seriously considering war. Wolzogen's plan involved a spoiling attack into Poland to devastate that area prior to Napoleon's expected arrival and as a way to implement a "forward defense." Wolzogen anticipated a fierce Napoleonic reaction and that Russian forces would then withdraw, bleeding off the strength of the French army with scorched-earth, guerillas, and a protracted fighting retreat as the Russian army withdrew into the vastness of Russia.[33]

By 1811 Wellington's success in Portugal was well known, but the adoption of a purely defensive strategy posed dangers to the Tsar that could result in his own deposition or even assassination in favor of one of the many Grand Dukes. Alexander had to tread carefully. Nonetheless, most of his advisors favored a defensive posture, assuming the time for a preemptive offensive against Poland had passed. The Tsar was also greatly influenced by his personal military advisor Ernst von Phull, another Prussian general. Phull, who had served with Scharnhorst on the pre-1806 Prussian

Quartermaster General Staff, envisioned creating a fortified camp à la Frederick the Great at Drissa (100 miles east of Vilna) that Barclay's army might fall back on. From here the Russians might accept battle in a position of strength with a defensive line along the Dvina River and control access to the line of march to the south over the Beresina River to the Dnieper River. None of the Russian planners—Phull, Barclay, or Wolzogen—anticipated retreating any further east than this "three rivers" line.[34] Additionally, Bagration's southern army, reinforced to 80,000 troops, could operate per the Wolzogen-Barclay plan against the French flank and rear—or if Napoleon turned on Bagration, Barclay could do the same. In short, the Russians intended to open up the campaign with a shallow strategic withdrawal and a great defensive battle, or even a flank attack against an isolated Napoleonic force. Recall that this last component had been Bennigsen's modus operandi during the Eylau-Friedland campaigns (see chapter 5), but now the Russians planned to use two armies instead of one to whittle down the French forces using attritional warfare. This optimistic plan fell apart once Bagration and Barclay realized the immense size of the forces facing them. Additionally, although Barclay was now 51 years old, he was younger than many more senior generals who resented his elevation over them. Worse, Barclay was not even Russian, but rather Livonian with some Scottish blood and a Baltic accent that was often regarded as "German"—the Russian euphemism for foreign officers. Bagration, too, was not Russian, but as a Georgian-Armenian regarded as more Russian than Barclay. Bagration also harbored an intense jealousy that Barclay had obtained the supreme command instead of him, one of Suvorov's protégés. Finally, in the wings waited the disgraced recipients of former Napoleonic drubbings, Prince Mikhail Kutusov and Count Bennigsen, ready to take command if Barclay (still the minister of war) faltered.[35]

The size and geographic scope of the upcoming campaign presented a difficult command-and-control challenge for both sides. As we have seen, Napoleon created a new echelon, the army group, with himself in command, to help solve this problem. This would have been fine had Napoleon trusted his subordinates and allowed them more decision-making initiative at the operational level, what is called mission command in the U.S. Army (see chapter 4). However, other than Eugene (and not all historians agree on this point), Napoleon had done little to develop the habit of independent command in his subordinates. Some had demonstrated this talent earlier in their careers (Masséna) or as a result of accident and moral courage (Davout).[36] Thus, although Napoleon created a de facto army group to help solve this situation, he had to command the army under his personal command, too (which was initially 222,000 men strong). Additionally, he treated Jerome, Eugene, Macdonald, and Schwarzenberg as he did the corps commanders under his direct control instead of trusting them (although perhaps he trusted Macdonald *too* much). Thus, he overburdened himself and his staff in trying to command and control at so many levels. Often his orders, if they arrived at all through the Cossack-infested country, arrived well after the situation had changed and only contributed to hesitation and confusion by his subordinates rather than adding value. The size of war had grown beyond the Emperor's means to effectively control all its many parts, despite his computer-like brain and astonishing work ethic. He was older now and his capacity for work had atrophied as had his physical condition since 1809. Napoleon's endurance had declined just as he began a campaign that would test him mentally and physically more than any other he had conducted.[37]

On the Russian side the problems tended toward the reverse—no one was in charge. This was largely Barclay's doing due to his publication, as minister of war, of a new regulation governing the command of armies that gave Russian generals virtual autonomy.[38] Alexander was nominally in command of everything at every level. Yet, instead of accompanying the army at all times, three weeks after the campaign started he departed the headquarters and generally remained behind the front lines, commanding at what might be called the political-military level, much as a U.S. president does today. However, he never appointed Barclay supreme commander and that officer had to use his position as war minister to try to direct Bagration. Although nominally under Barclay's orders, Bagration often ignored them. Tormassov, Tshitshagov, and Steinheil took their marching orders directly from the Tsar. When the Tsar made recommendations, they were often just that. His commanders were more ready than most of Napoleon's to make decisions based on the actual situation on the ground or in their sector and politely ignore operational guidance from the Tsar. Ironically, this loyal insubordination led to a far more effective system of command on the Russian side of the hill that more closely matches mission command than did Napoleon's system.[39]

The first problem for Barclay and Bagration turned out to be one of timing. Napoleon thrust across the Niemen on June 24–25, getting over 120,000 troops across that river in two days, and advancing rapidly toward Vilna. The other units of his army crossed the river at other points, although both Jerome and Eugene were delayed by poor roads and lack of orders, respectively. Also, Bagration's army was well short of the 80,000 troops envisaged by Phull's plan and, worse, dangerously separated from Barclay. This separation exposed the Russian armies to defeat in detail. Because of the Drissa axis of retreat, they fell back on axes that diverged from each other with Davout's extremely large I Corps (70,000 men) advancing between them. Murmuring soon arose among the native Russian generals as Barclay retreated without a major engagement toward Drissa. It soon became apparent that Drissa and several supporting fortifications (e.g., Dunaberg) were far from ready. No less an authority than Carl von Clausewitz, who had resigned his commission in the Prussian army and now served in the headquarters of Barclay's army, assessed Drissa as follows: "if the Russians had not voluntarily abandoned this position, they would

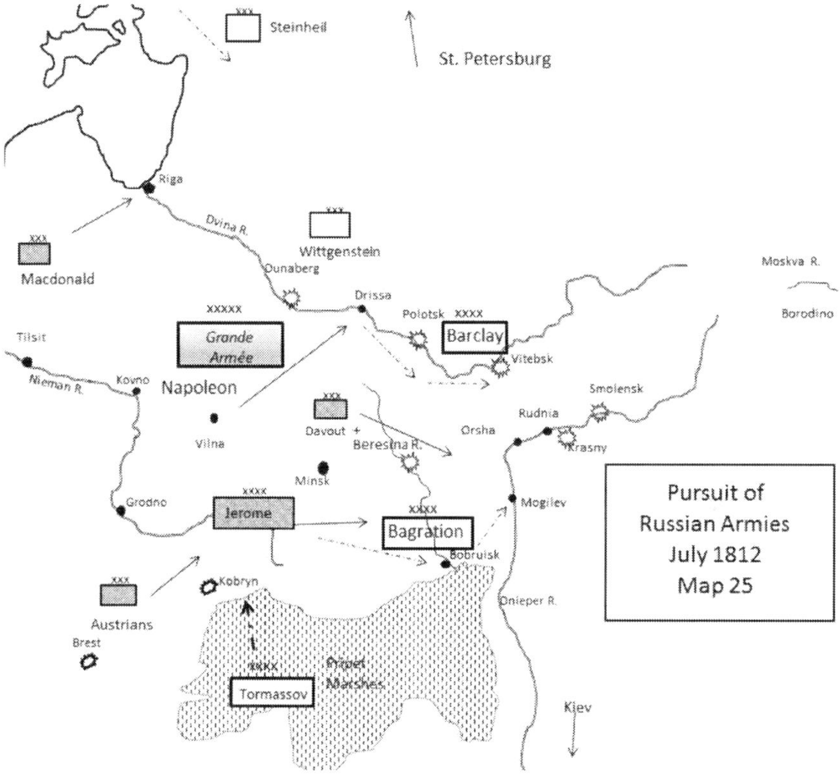

have been attacked . . . driven into the semi-circle of trenches and forced to capitulate."[40] Arriving at Drissa on July 8 ahead of the army, Barclay realized the position's unsuitability. Accordingly, he made several excellent operational decisions. First, he detached the large corps of General Count Ludwig Wittgenstein (25,000) to operate independently northeast of the Dvina river line, protecting St. Petersburg and threatening the French flank. Secondly, he retreated to the southeast toward Vitebsk (July 18), correcting the error of the divergent axis with Bagration by moving toward that general's line of retreat (see map 25).[41]

The retreat from Vilna by the First Army toward Drissa meant that Napoleon's first blows hit relatively empty terrain, made emptier by the Russian scorched-earth policy and inhospitable due to roving Cossacks belonging to the Hetman Platov. Platov's Cossack Corps had the job of both covering the retreat and linking the armies of Barclay and Bagration. In the upcoming weeks, these nimble Cossacks, sometimes supported by regular Russian cavalry, usually bested their French counterparts in the numerous cavalry skirmishes of the campaign. Unable to catch Barclay at Vilna,

Napoleon adopted the second course of action mentioned before; he decided to risk outstripping his plodding supply trains (and boats) and press on.[42]

The vulnerability of Bagration's Second Army now attracted the Emperor's primary attention as he sent Davout south to cooperate with Jerome in trying to destroy it. However, Jerome moved slowly and assumed Davout was under his command in this ad hoc army group, not Napoleon's. Further, Davout had almost as far to march to reach Bagration as the Russian did to escape him. Finally, the confusion of command and ignorance about Bagration's anticipated line of retreat resulted in Davout failing to trap Bagration, although this maneuver temporarily prevented Bagration's juncture with Barclay. Napoleon exploded in wrath against his younger brother, sending a letter so insulting that when Jerome received it on July 14 he turned over his command to his chief of staff and promptly returned to Westphalia three days later, taking his Westphalian Royal Guard with him. Ironically this was the same day that Barclay began his southeastward movement away from Drissa toward Bagration.[43]

Napoleon's huge army found itself spread across almost 300 miles of frontage from Riga to the northern edge of the Pripet Marshes. His lumbering juggernaut also suffered immensely from straggling and the heat of the Russian summer. Poorly cared-for soldiers and animals littered the crude roads of Lithuania, Poland, and White Russia (Byelorussia). Cossacks and peasants alike (once in Russia) murdered or took them prisoner, usually the former. Water, food, and rest were all in short supply. The wastage of draught animals and horses in the first two weeks of the campaign was beyond anything anticipated by Napoleon and was made worse by the failure of the French cavalry to maintain a good picture of Russian movements. His cavalry included many new horsemen who had very little training on the care and feeding of horses, further decimating the mounts of the cavalry.[44] Additionally, unseasonable heavy rains now began to alternate with the heat and combined to make more men, horses, and oxen sick and further slowed movement by turning the roads to mud. Horses foraged not on prepared fodder (oats and barley) but on the unripe crops of this season, which added digestive ailments to the other causes of death. Around 10,000 dead horses alone littered the roads to Vilna, causing Napoleon to leave 80 guns there for lack of transport. The Russian summer of 1812 is not given its due in many histories of this campaign, but its impact in the opening months was as significant as the effects of the more famous winter later that year.[45]

Bagration, in the meantime, regarded himself as sorely used by his "non-Russian" nominal superior in command. As Barclay withdrew toward Drissa, the Tsar had dispatched orders for Bagration to come north to unite with First Army. This Bagration had done, running into the advancing French and then turning south again; deciding he could not break through the entire French army, which he believed was at Minsk. He blamed all his misfortunes on Barclay, believing that Napoleon himself had come south with his entire army to destroy him personally (Napoleon had remained in Vilna). Despite his bravado and accusations of treason against Barclay, he wisely decided to retreat due east through Bobruisk, turning north to join Barclay at Vitebsk by way of Mogilev.[46]

Barclay, now in sole command of the First Army after the Tsar's departure, hoped to unite with Bagration near Vitebsk and there fight the battle the Russian generals and the Tsar clamored for. Responding to his understanding of the situation, Napoleon left Davout in charge of the pursuit of Bagration and concentrated the bulk of the *Grande Armée* under himself. However, in doing so he made another significant detachment from his forces. The Saxon Corps (VII) under General Reynier moved south to protect his strategic rear in Poland from Tormassov's army. Napoleon now tried to correct the mistake he had made in putting Schwarzenberg's Austrians on the far right flank, ordering them to come north and join the army group in the center.[47] On the northern flank Napoleon had committed himself to a grand operational maneuver that has become known to history as "the maneuver on Vitebsk."[48] His goal was to use the forces on his left (northern) flank to engage Barclay on the right bank of the Dvina while he took the main army to Vitebsk and cut off the Russians from both Smolensk and Bagration. The problem with this plan was that it was implemented without accurate knowledge of where Barclay actually was and relied heavily on Murat's cavalry reserve to cause the First Army to slow down and fight rearguard actions. The only actions being fought were between the two cavalry forces, with Murat often bested and the Russian guns and infantry reaching Vitebsk before Napoleon. After arriving at Vitebsk, Barclay came close to making an error in bowing to the will of the Russian hotheads the Tsar had left behind and fighting a battle. However, circumstances conspired to allow the Russians to escape as Barclay waited for Doctorov's Corps and Pahlen's cavalry to rejoin the First Army for the battle. To cover these troops' arrival, Barclay ordered his rearguard to fight delaying actions all day (July 25) in advance of his position with other detached troops. After Doctorov and Pahlen arrived, Barclay received Bagration's message that he had been blocked by Davout from coming north near Mogilev on July 23. Bagration was much criticized for not plowing through Davout, which he outnumbered 2–1, but his decision proved fortuitous. This news convinced Barclay to withdraw and this decision in turn denied Napoleon his best opportunity to destroy one of the Russian armies before they united. By July 27, Napoleon had gathered more than 135,000 troops versus Barclay's 75,000—a French victory was almost certain had Barclay stayed put.[49]

Nonetheless, on July 26–27, Barclay's rearguard again fought aggressive rearguard actions to trick Napoleon about his intentions and then made good his escape to Smolensk, arriving there on August 2. Bagration arrived the following day, having countermarched down the Dnieper and crossed that river before pushing northeast to Smolensk. Napoleon's attempt to keep the Russian armies separated had failed, although some historians claim he let them unite in order to destroy them at one blow, giving him a prescience he did not have. At any rate, after his army group's three exhausting maneuvers on Vilna, Drissa, and now Vitebsk, Napoleon called a halt to the pursuit and gave his troops a rare week-long operational pause. He even considered, briefly, ending the campaign for that year.[50] Napoleon, despite his first campaign in Italy and the experience of Spain, did not believe in protracted war. As he told his marshals, "Nothing is more dangerous to us than a prolonged war."[51] Historian Richard Riehn captured a reasonable operational explanation for why Napoleon continued a campaign that had already seen one-fourth of his first echelon army group melt away to desertion, disease, and exhaustion:

> Despite the cost of the campaign so far, it still had not occurred to him that his method of campaigning was out of place in this theater of war and that it was . . . the most important source of his losses. Yet given the . . . expanses of Russia . . . [and] the means of communication, success, if it was to be had at all, could come only through a *systematic* conduct of the war. Just as the exploitation of [Russia's] resources would require much time . . . so would the securing of these require *a step-by-step advance, establishing new operations bases* because no territory could be left behind with its resources not secured. This kind of warfare . . . was totally alien to Napoleon's style.[52] [emphasis added]

Napoleon the operational artist did not adjust to the imperatives of a protracted campaign as he had done earlier in his career in both Italy and Poland, relying instead on an Austerlitz or Wagram masterstroke to solve all his strategic problems. When that stroke did not materialize he seems to have simply "doubled down" to try and press on for another roll of the dice and a decisive battle. He no longer took *risks*, rather he began to *gamble* with the very success of the entire war.

Worse, as Napoleon rested his troops at Vitebsk, trouble developed on his flanks. In the south his attempt to have Reynier's small corps of 17,000 Saxons protect Poland from Tormassov's army of over 40,000 received a rude shock when Tormassov seized Brest and then destroyed one of Reynier's brigades at Kobryn on July 27 (see map 26). If Tormassov had acted more aggressively he might have bagged Reynier's entire corps; as it was, Reynier retreated and sent out a message for help to Prince Schwarzenberg, who turned around with his strong Austrian corps of 30,000 to support him. Napoleon now directed Schwarzenberg to take command of both corps. In the north Marshal Oudinot crossed the Dvina to attack Wittgenstein north of Polotsk. Wittgenstein performed his role as an operational economy of force to perfection. He first reinforced his corps with the garrison of Dunaberg, abandoning that fortress, and then bested Oudinot in a series of actions in late July and early August. By August 2 Oudinot was back in Polotsk, requesting that Napoleon send him reinforcements to keep Wittgenstein at bay. Napoleon sent the weakened Bavarian Corps (13,000 men) to help. It was commanded by General Gouvion St. Cyr, one of the

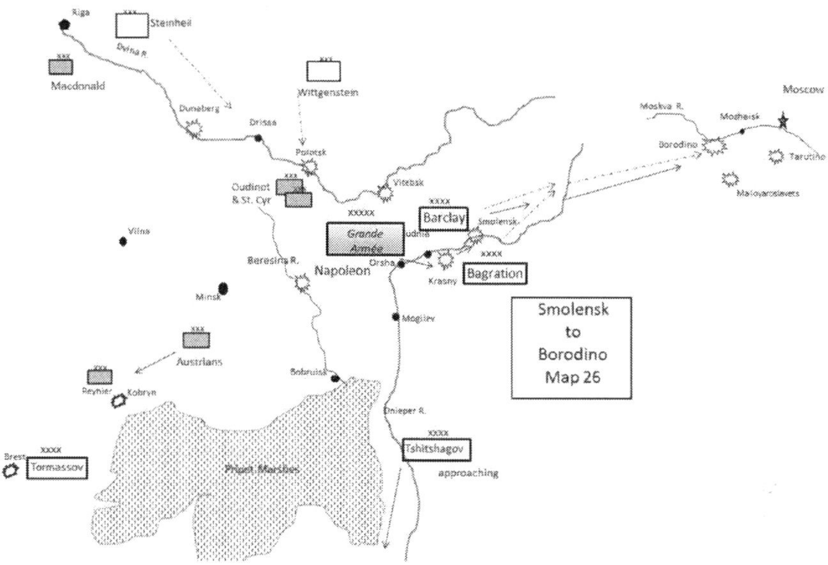

few generals in Napoleon's army capable of independent command, but St. Cyr, for the moment, remained subordinate to the hapless Oudinot.[53]

These actions now allow us to compare the use and efficacy of both sides' operational reserves for the first five weeks of the campaign. In Napoleon's case his secondary operational echelon had been whittled down by three entire corps to deal with Russian flanking forces. Barclay detached one corps but Napoleon had to eventually detach two to account for it (not including Macdonald). Worse, Wittgenstein won several actions, further increasing Russian morale and momentum and demoralizing the French. On the southern flank, Napoleon had attempted to bring up Schwarzenberg's strong and relatively healthy Austrian corps for use as a fresh operational echelon. Instead, because he left too weak a force (Reynier) to cover his lines of communication in Poland, Napoleon was forced to send this corps back.

As for the Russians, they had suffered grievously too, but not as badly as the French, losing 30,000 troops to all causes as compared with over 100,000 for Napoleon. While Napoleon's reserves struggled to catch up or were detached to deal with crises on the flanks, the Russians collapsed on their own lines of communication, integrating 20,000 reserve troops as they retreated. While the Russians fell back on fresh reserves waiting at well-stocked depots, the French replacements and reserves made exhausting marches in the hot sun to catch up with the leading elements of Napoleon's always-advancing army group. Finally, the Russians proved, as they had in Italy in 1799, that they could keep pace with, and even outmarch, the French infantry. Because their forces were smaller and more disciplined they lost fewer stragglers and often marched at night to avoid the heat of the day. Further, Alexander had ordered the mobilization of militia after returning to Moscow and then directed another round of conscription in mid-August. Fresh, albeit raw and untrained, troops would join the Russian army in growing numbers the longer the campaign proceeded.[54] Napoleon's margin of superiority grew more slender with each kilometer that he advanced, despite the immense resources and care he had taken to ensure that on the day of battle his superiority would be decisive.

Bagration, after a cordial meeting with Barclay, had agreed to subordinate himself to Barclay, but he had also animated the latter with his bellicose spirit. Barclay yielded and decided to attempt an offensive—a spoiling attack much like Wittgenstein's—against Napoleon's forces advancing lethargically (according to Russian intelligence) from Vitebsk. This idea was founded on the notion that Napoleon was far weaker than he actually was and had been buttressed by the news of the successes on the French flanks.[55] In reality, Napoleon's main army group approached 190,000 men while Barclay and Bagration had a little over 120,000. The Russians were about to offer Napoleon what he most craved—a large battle.[56] Napoleon's plan remained the same, to hold on his left and attempt to march around the Russian left flank and rear at Smolensk. To do this he would have to march farther than if he advanced directly toward Smolensk from Vitebsk. Clausewitz wrote of this decision, "It is inconceivable how Bonaparte omitted to throw forward his right wing, so as to cut off the Russians from [the direct road through Rudnia]." In hindsight, Clausewitz was right; had Napoleon gone straight at his objective, the united Russian army, he would almost certainly have engaged it in a significant meeting engagement north and west of the Dnieper, although whether he would have destroyed it is open to debate.[57] Both forces hit relatively empty terrain, like two blind boxers, although Barclay's advance guard surprised and defeated a French cavalry division near Inkovo on August 8. This action caused Napoleon to move aggressively against Smolensk. As the Russians advanced north of the Dnieper west of Smolensk, Napoleon's main body swung around near Orsha and crossed the great west to south bend in the Dnieper River. However, Barclay's stop-and-start offensive halted for good on August 12, when, in Clausewitz's quaint phrase, "he became anxious about his retreat."[58]

Napoleon's advance guard under Murat—three cavalry corps numbering almost 15,000 troopers—arrived at Krasny on August 14, where the dilatory Bagration, operating on Barclay's left, had placed a division under General Neverovsky. The Russians were hustled out of Krasny, fighting desperately against Murat as they withdrew in a divisional square toward Smolensk. Fortunately, the slow speed of Bagration's advance, in part due to his concern over the open left flank at Smolensk, allowed him to throw General Raievsky's nearby corps (17,000) into Smolensk, preventing Murat and his supporting infantry from capturing it outright on the following day. The result was a confusing series of battles around Smolensk. Neverovsky notified the "anxious" Barclay of his misfortune on August 15 and that was all that was needed for Barclay to make his decision to continue the retreat. Thus it was that the two character flaws in the main Russian generals, irresolution for Barclay and insubordination for Bagration, combined serendipitously to save the Russian army, although much bloody fighting had to be done to do it. Napoleon fought an indecisive series of battles from August 15–19 around Smolensk. Pummeled in the early days of these engagements, Bagration's army escaped to the east on August 17, Barclay, fighting several stiff rearguard actions under very trying circumstances, followed on August 19. Clausewitz dryly noted that each side lost around "20,000 men" although he includes the number from Krasny in his totals. Smolensk, proud city of Old Russia, lay in smoldering ruins.[59]

Barclay's actions from August 3–20, through luck, occasional resolve, and "anxiety," saved the Russian army, but it doomed his position as the de facto commander of the Russian army group. Russian generals and favorites of the Tsar, who felt he had denied them a wonderful victory, convinced Alexander to dismiss Barclay from command. As the battles around Smolensk raged, the Tsar reluctantly convened a committee to pick a replacement for Barclay as supreme commander on August 17. Prince Kutusov got the nod as commander with the scheming regicide Count Bennigsen as his chief of staff. Barclay, instead of being drummed from the army, was demoted back into his original operational billet as commander of the First Army. Kutusov joined the army as Barclay continued to withdraw toward Moscow on August 29. Kutusov faced a tricky situation since, like Barclay, his inclination was to continue to withdraw. However, the Tsar and most of the Russian army expected him to halt and fight Napoleon. Kutusov knew that if he did not fight the back-stabbing Bennigsen might take his place.[60]

As the Russian army had withdrawn, it absorbed another corps of replacements under General Mikhail Miloradovitch, nearly making up the army's losses at Smolensk. Napoleon was not so fortunate, although on his flanks both Schwarzenberg and St. Cyr (who took command on August 17 from the wounded Oudinot) had won victories that removed, for the time being, the threats to his rear. These victories, as it turned out, gave Napoleon a false sense of security about his strategic flanks. Napoleon's main army group was now down by 20,000 and by the time he reached Borodino his margin of superiority over the Russians had dropped to approximately 134,000 troops versus 120,000 Russians. However, 10,000 of the Russians were the recently mobilized militia, many armed only with pikes. Napoleon's greatest fear was that the Russian army would deny him a battle and again retreat, but Kutusov did not disappoint him.[61]

The first elements of the French advance guard arrived near the little town of Borodino just west of the confluence of the Kalatsha and Moskva (Moscow) Rivers on September 5, fighting a stiff battle against Bagration's rearguard. All this did was to subtract another 4,000 troops from each side. Napoleon then brought up the rest of his army on September 6 and on the following day the holocaust of Borodino occurred. Only the sea battle of Lepanto (1571) exceeded this bloodletting in terms of a single day's slaughter until the disastrous first day of the Somme in 1916. Some accounts list a combined casualty count of over 100,000 with as many as 40,000 killed, although a total of 80,000 casualties for the two armies is probably closest, with the Russians losing almost twice as many as the French. Most were lost to mass artillery fire, with 587 guns on the French side and 624 on the Russian, spitting canister and round shot nonstop for most of the day.[62] It was the first battle where two army groups organized as such fought head to head.

Ironically, Kutusov had left the actual conduct of the battle to Barclay, Bagration, and Platov, his principal subordinate commanders. Its final phases were almost entirely under Barclay's tactical control after Bagration was mortally wounded. By the time the fighting ended at 4:00 P.M. it was Barclay who stabilized the front, keeping the mauled Russian army on the field. After the battle, Barclay was cheered by the troops wherever he went. Kutusov, meanwhile, sent a message to the Tsar declaring victory and went to bed. He was roused in the middle of the night by reports of the loss of almost half the army and he gave orders to pull back several miles to Mozhaisk. The French army, also badly damaged, withdrew in some places for the night, convinced of a victory, and then apparently confirmed as much the next morning when it found the Russians retreating in three columns. Murat asked Napoleon to release the Imperial Guard cavalry for a pursuit but he was denied permission. Napoleon has been much criticized (although, oddly, not by Clausewitz) for his failure to release his last tactical and operational reserves both on the day of battle and afterward to seal his victory.[63]

Borodino solved little for either side. Kutusov had fought the battle everyone had clamored for, lost half the army, and now his goal was to save the other half. He continued his retreat, abandoning 10,000 wounded in Mozhaisk, while boldly proclaiming his plan to fight another battle before Moscow, but this was designed to obscure his real intent. When a council of war (including Bennigsen) recommended another battle, he ignored them, making his most important command decision of the campaign and leaving Moscow to the *Grande Armée*. Clausewitz implied that this was the most important Russian operational decision of the war. Credit for the eventual victory, which was not so clear at the time, goes to both Barclay and Kutusov: the former preserved the army at Borodino and the latter preserved it afterward.[64]

THE RETREAT

Napoleon now made his final disastrous error of the campaign. Instead of making the Russian army his objective, as he had so often in his career, he settled for the occupation of Moscow, holding it hostage in anticipation of a negotiated peace. Kutusov retreated to the southwest of Moscow, just out of reach of the French in unforaged countryside undamaged by war. This location allowed him, in Clausewitz's phrase, to "play hide-and-seek" in the vast expanses of Russia should Napoleon come after him.[65] It was also nearer to the armies of Tshitshagov and Tormassov. Napoleon had passed the "culminating point of victory," but it took five weeks of intolerable indecision as Moscow burned around him to recognize this fact. It was at this point that Napoleon's decisions no longer influenced the issue of retreat; it was foreordained when the Tsar refused all emissary attempts in St. Petersburg. The issue now to be decided was the scale of the disaster and Napoleon's long halt in Moscow during the last good weather of the year ensured it would be catastrophic.[66]

The remainder of the campaign has often been identified as the cause of Napoleon's defeat: a long retreat through the bitter winter that destroyed the *Grande Armée*. In contrast, the evidence presented suggests that army was already defeated when it marched into Moscow and doomed as it marched out. Kutuzov's army slowly increased in size until it outnumbered Napoleon's—by October it numbered approximately 100,000 troops. Napoleon's operational reserves were strung out on his flanks and lines of communication and his army got no stronger as it sat idly in Moscow. Napoleon received nothing but bad news from his flanks, and despite making St. Cyr

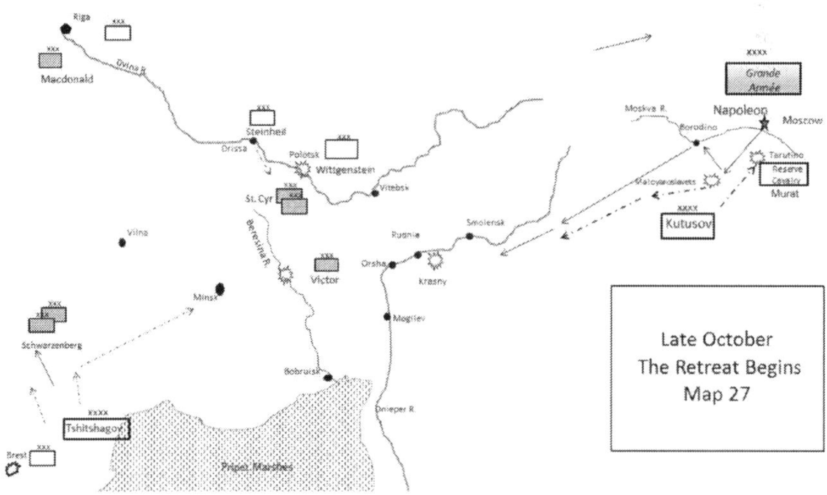

Late October
The Retreat Begins
Map 27

a marshal after that general's victory at Polotsk (August 18) against Wittgenstein, the arrival of Steinheil's Finnish Corps in combination with Wittgenstein caused the wounded St. Cyr to retreat. In the south, Tshitshagov, now in overall command, sidestepped Schwarzenberg and began advancing north against Napoleon's lines of communication. Kutusov contented himself with bringing Tormassov up from the south to replace Bagration (who had died of his wounds) and waited on events. Sensing the time was right, he surprised Murat's covering force south of Moscow at Tarutino on October 18 (see map 27). Murat lost nearly 4,000 men and 36 guns. This defeat convinced Napoleon to retreat and Moscow was evacuated on October 19 in relatively good weather. The *Grande Armée* numbered no more than 95,000 troops.[67]

Napoleon's intent was to do what the Russians had done and incorporate his reserves as he retreated and then to winter in Smolensk if circumstances permitted it. His route of march to the southwest through relatively unravaged Russian countryside had much to recommend it and might get him to Smolensk, if all went well, by early November. Autumn rains slowed him down and on October 24 at Maloyaroslavets the advance guard of his army under Eugene collided with Kutuzov's main army, giving both forces a rude shock. Napoleon, not realizing that Kutusov had pulled back and opened the escape route west, countermarched, losing another week of reasonably good weather, and retreated to the more northerly Moscow-Smolensk road, scene of so much death and misery that summer. Winter arrived as the first snows fell on October 31. Kutusov marched parallel to the French line of retreat, harrying their every step with Cossacks. Without the glue of promised victory, Napoleon's army began to dissolve. His army was in such bad condition by the time it arrived in Smolensk (November 9) that it had fewer than 30,000 effectives and another 20,000 stragglers. Worse, Kutusov cut his line of retreat at Krasny and Napoleon was forced to use his Imperial Guard on November 16 to cut his way through. Marshal Ney's corps, bringing up the rear was cut off. Ney, retreating north and then south again, managed to bring a pitiful 900 men back into the army, but Krasny was a major disaster. In addition to 6,000 casualties, some 26,000 French stragglers were captured.[68]

Where were Napoleon's operational reserves, including the supposedly fresh corps of Victor? These forces were fighting for their lives and also retreating. Wittgenstein and Tshitshagov advanced from the north and south, respectively, toward the Beresina River to cut off Napoleon's retreat (see map 28). Here a brief flash of Napoleonic brilliance occurred as he defeated Tshitshagov and bluffed Wittgenstein from 25 to 29 November. Kutusov, now seriously ill (he would die in the spring of 1813), hung back with the main army, satisfied to let Napoleon leave Russia without risking a defeat. Victor's still-intact corps and Oudinot's bloodied one united with the main army and did most of the fighting. However, as at Krasny, once the disciplined core of what was left of Napoleon's "army group" crossed the Beresina, tens of thousands of stragglers were cut off and fell into enemy hands. According to Clausewitz: "Chance

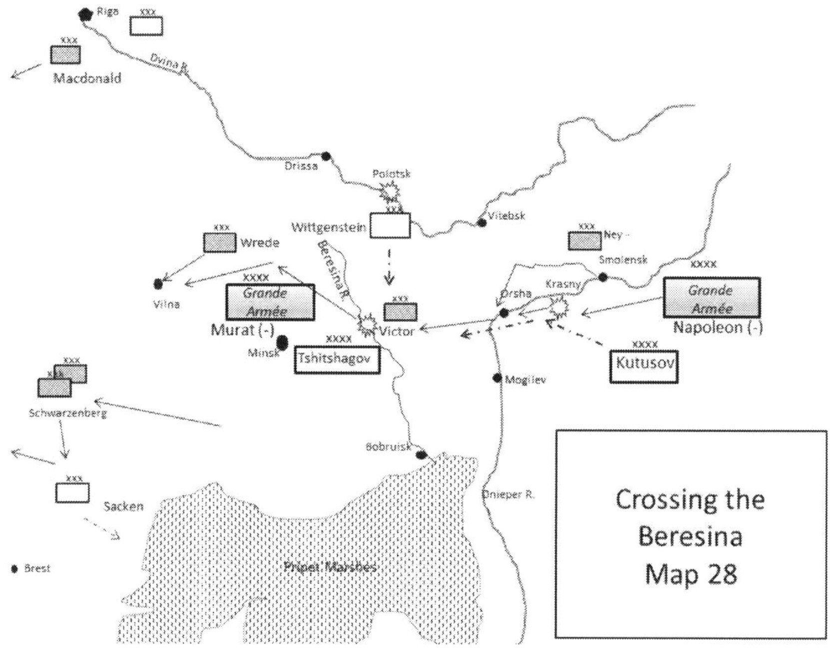

certainly somewhat favored Bonaparte . . . but it was his reputation which chiefly saved him, and he traded in this instance on a capital amassed long before."[69] Some accounts list as many as 30,000 stragglers—no longer fighting but struggling to avoid dying of starvation, exposure, or both—falling into Russian hands. David Chandler estimated the losses as coming to perhaps 50,000 to 60,000, including many who drowned or froze in the frigid river. Napoleon stayed with the army until December 5 and then departed at Smorgoni for Paris to quell conspiracies, propagate the myth that the Russian winter had defeated him, and prepare for the worst, turning over what was left to Murat.[70]

* * *

Napoleon's army, including replacements near Vilna, lost another 40,000 to 50,000 troops as it withdrew from Russia—many of them second and third echelon replacements who succumbed to the bitter cold and died or were captured. Of the 66,000 troops that remained in Poland by mid-December, nearly 43,000 were Austrians and Prussians who soon deserted Napoleon (see chapter 8). He now fell back on what might be metaphorically termed his *strategic seed corn*. The cadre of surviving French soldiers, probably no more than 2,000 of them officers, formed the nucleus for Napoleon's armies of 1813, but he had sacrificed his strategic reserves in Russia and the rest remained tied down in Spain in a war they had already lost. His cavalry would never recover from its Russian losses and he lost more artillery in this campaign than any general in history. The campaign had also seriously damaged the Russians, Kutusov would soon die and the army fell into the hands of Wittgenstein. While Napoleon probably lost over half a million men in Russia, the Tsar's legions suffered nearly a quarter million losses. Alexander's protracted war of attrition had ejected the conqueror from his nation and circumstances soon made it possible for him to continue hostilities into Germany.[71] We leave the last word to Clausewitz, who spent the majority of the campaign in the headquarters of the main Russian army:

> Bonaparte determined to conduct and terminate the war in Russia as he had so many others. To begin with decisive battles, and to profit by their advantages . . . to go on playing double or quits till he broke the bank . . . In Spain it had failed. . . . It is extraordinary, and perhaps the greatest error he ever committed, that he did not visit the Peninsula in person. . . . He reached Moscow with 90,000 men, he should have reached it with 200,000. This would have been possible if he had handled his army with more care and forbearance.[72]

EIGHT

OPERATIONAL ART AND THE CAMPAIGN OF 1813

CHRONOLOGY

Dec 30, 1812	Convention of Taurrogen, Yorck neutralized his Prussian Corps.
Jan 30, 1813	Austria secretly declared neutrality.
Mar 1, 1813	Treaty of Kalisch: Prussians joined the Russians to create the Sixth Coalition.
May 2	Napoleon won battle of Lützen/Gross-Görschen.
May 20–21	Napoleon won battle of Bautzen.
Jun 4	Napoleon agreed to cease-fire at Pleischwitz.
Jun 21	Wellington defeated Joseph Bonaparte at Vitoria.
Jul 9–12	Allies met at Trachenberg.
Jul 25–30	Wellington defeated Soult's counteroffensive in the Pyrenees.
Aug 13	Austria declared war on France.
Aug 15	Austria, Prussia, Russia, and Sweden resumed active operations against Napoleon.
Aug 23	Marshal Oudinot was defeated at Grossbeeren.
Aug 26–27	Napoleon defeated Allies at Dresden.
Aug 26	Blücher defeated Marshal Macdonald at the Katzbach.
Aug 29–30	Vandamme's French corps annihilated at Kulm.

Sep 6	Bülow defeated Marshal Ney at Dennewitz.
Oct 15–19	Napoleon was defeated at Leipzig.
Oct 30	Napoleon defeated Bavarians at Hanau and retreated into France.

On December 30, 1812, General Hans David Yorck, commander of the Prussian component of Marshal MacDonald's corps completed extensive negotiations with a Prussian in Russian service—Carl von Clausewitz. Yorck agreed to neutralize the 17,000 Prussian troops under his command for a period of two months. With this agreement, signed at Taurrogen in Lithuania (not far from Tilsit), Clausewitz effectively inaugurated the Sixth Coalition against the Empire of France (see map 24). The Convention of Taurrogen served as the catalyst the Russian Tsar Alexander had hoped for to continue his holy crusade against Napoleon, who had despoiled his nation the previous year.[1] The outgrowth of this event was the subsequent rebellion of Prussia against Napoleon in early 1813. For the first time since 1807, two major continental powers, Prussia and Russia, were united in arms against Napoleon's Empire. In the spring operations that followed, Napoleon emerged tactically victorious despite the military improvements of his opponents, the weaknesses of the new *Grande Armée*, the wavering of Napoleon's allies in the Confederation of the Rhine, and the unexpected (to Napoleon) neutrality of Austria.[2]

The campaign by the Sixth Coalition against Napoleon in 1813 in Germany offers a particularly lucrative case study by virtue of its sheer scope of operations. The coalition's operations in Germany were only one of many theaters with a strategic bearing on the situation; but the liberation of Germany was by far the most important strategic goal when compared to the liberation of Spain and Italy (the latter still not completely "liberated" at the time of Napoleon's abdication in 1814).[3] It was in Germany in 1813 that we see the employment of the mass *coalition-conscript* armies by the Allies—Prussian, Austrian, and even Russian. We have seen the improvements effected by all the nations that comprised that coalition save one—Prussia. It is to that nation that we now turn before continuing to examine operational art in 1813.

PRUSSIAN REFORMS

Like the Russians and Austrians, the Prussians implemented reforms in response to the threat posed by Napoleon. The course of these reforms was by no means a smooth one for Prussia—in the interim Napoleon reduced her to a second-rank power by constraining the size of her army to approximately 42,000 troops total.[4] The Prussians, however, became a true nation in arms in the manner of France and came up with innovations of their own, perhaps the greatest being their general staff.[5] Of all the opponents of France the Prussians were probably the ones who most completely and competently absorbed the new ways of warfare.[6] The visceral incorporation of these methods had already begun prior to the Prussian catastrophe of 1806, which was to rapidly accelerate an already established reform movement within the Prussian army. Its genesis was, as with the Austrians, the defeats of the early Revolutionary Wars. In 1795, after Prussia withdrew from the First Coalition, the War Ministry in Berlin established a commission to "investigate and ameliorate the defects that had appeared" during the recent conflict with France. Aside from an increase in light infantry (termed fusiliers or jaeger by the Prussians) nothing of real importance was accomplished.[7] Two events then occurred that helped the Prussian reform movement gain momentum: a new king interested in military affairs was crowned in 1797 (Frederick William III), and Gerhard von Scharnhorst was accepted into the Prussian army for service as a quartermaster general (chief of staff).[8]

According to Scharnhorst's protégé, Baron Karl von Müffling, "Scharnhorst had made Napoleon's mode of warfare, and the means of resisting him, the chief object of his study, and endeavored accordingly to prepare young men for the war . . . with this dangerous opponent."[9] The catastrophe of 1806–1807 provided the reformers, led by Scharnhorst, with the opportunity to completely revamp the Prussian army. Clausewitz summarized the gist of Scharnhorst's reforms:

1. Weapons, equipment, and an organization consonant with the new methods of warfare.
2. To improve the individuals making up the army, and to raise their morale and dedication; consequently the abolition of foreign recruitment, steps toward universal service and the abolition of corporal punishment, and the establishment of good military schools.
3. An appropriate system of promotion to supplement promotion by seniority alone, and the careful selection of the commanders of larger units.
4. New training methods appropriate to modern warfare.[10]

These recommendations were adopted along with a purge of nearly 5,000 officers.[11] Finally, Scharnhorst implemented a policy of pairing battle tested commanders such as Yorck and Blücher with professionally trained staffs headed by himself (and after his death Gneisenau) and the cream of the *Kriegsakademie* (war academy)—Hermann von Boyen, Müffling, and Clausewitz. This resulted in an operational command system second to none by 1813.[12]

Scharnhorst's response to Napoleon's limit on the size of the Prussian army involved an ingenious system of training that created a trained reserve—the *Krümpersystem*. During every training cycle, trained cadre were released (but kept track of) and new recruits called up to replace them. Gneisenau estimates that after three years as many as 150,000 men had been trained in this way.[13] However, only some 16,000 of these would be young enough for recall in 1813. The beauty of the system was that the ineligible trained men could serve in the landwehr (trained militia) that the reformers planned for in a national emergency. Additionally, the 16,000 trained reserves that were called up to service became cadres for other regiments and this resulted in an overall increase of around 40,000 additional trained troops to the Prussian army in early 1813.[14]

The published tactical doctrine for this army, finalized in the 1812 *Reglement*, earned the praise of Frederick Engels as "the best in the world."[15] A multiauthored document developed by Scharnhorst and Gneisenau, among others, its final form was due principally to an officer often regarded as a reactionary, Hans David Yorck. Napoleon's invasion of Russia served as the final element that honed the Prussians for the campaign of 1813. When Napoleon demanded that Prussia provide an auxiliary corps to support his invasion, many of its best officers, including Clausewitz, left the Prussian service for the Russian and thus gained valuable experience in fighting the French. Gneisenau reported that the efforts to rebuild the Prussian army via the *Krümpersystem* were in effect "paralysed [*sic*]." Nevertheless the final verdict of Gneisenau was that the loss of 10,000 trained men in Russia was "outweighed by the military experience gained."[16]

In summary, the Prussian army was now a trained all-arms "army of the people," albeit small, but with an excellent command system. Its "brigades" were roughly equivalent to the French divisions in size and superbly trained with a thoroughly up-to-date tactical doctrine. In turn these brigades were amalgamated into combined arms corps, each with an operational staff supporting the commanding general with separate artillery and cavalry reserves at the corps level. This army, thanks to British supplies would soon be excellently equipped in 1813 and was capable of expanding from about 30,000 veterans in January of 1813 to some 80,000 trained troops by the time of the spring campaign.[17]

OPERATIONAL ART IN THE SPRING 1813 CAMPAIGN

Yorck's defection at Taurrogen had far-reaching effects. Because he had neutralized his corps in Prussian Lithuania, Wittgenstein's leading elements crossed the Niemen River, effectively invading the Napoleonic domain. Shortly thereafter Yorck abandoned his neutrality and joined Wittgenstein, leaving Clausewitz among others to organize the landwehr and landsturm as well as replacement conscripts and volunteers for the regular regiments.[18] Subsequent operations can be divided into two distinct phases. The first phase began with Yorck's defection at Taurrogen and ended approximately at the time that Napoleon rejoined his reconstituted *Grande Armée* (named the Army of the Main) on April 25 at Erfurt.[19] The second phase began with Napoleon's active campaigning and encompasses the battles of Lützen and Bautzen, terminating with the signing of a cease-fire at Pleischwitz on June 4. This cease-fire was soon followed by a formal armistice.

Overlapping these two phases were other military operations that had a bearing on the operational situation. The first of these encompassed the masking and sieges of the fortified cities (e.g., Danzig) that Napoleon left along the Allied lines of communication. Other operations consisted of Allied attacks against Napoleon's lines of communications using partisans, friekorps, Cossacks, regular units, and the various combinations of all of these—in Soviet terms *deep operations*. U.S. Army doctrine from the 1990s defined deep operations as those that "engage enemy forces throughout the depth of the battle area and achieve decisive results rapidly."[20] Because of the vast geographic extent of these operations—one might call them distributed—their bearing on the campaign will also be examined.

Wittgenstein's invasion after Taurrogen completely changed the strategic fabric of the Russia's war against Napoleon. Until this point the war for the Russians was a *defensive* one. Its strategic aim was, in Clausewitzian terms, negative—expel the invaders from the sacred soil of Holy Russia. This was where the campaign would have reasonably ended. Let us first examine the situation as it appeared immediately prior to Taurrogen and then after to gain the context of the strategic and operational decisions that followed. For the Russians the year of 1812 had been more than just a French military catastrophe. One account estimates that the Russians had lost at least a quarter of a million soldiers killed alone. Because of these losses the Russians had just over 150,000 men left, including reserves and replacements, with only about 62,000, including Cossacks, available to carry the fight into Napoleonic domains. Additionally, Russia's military commander Prince Kutuzov, whose prestige was now immense, opposed a continuation of the pursuit of the French beyond Russia's borders (although it included the Grand Duchy of Warsaw). Russia had historically experienced great difficulties in projecting and supplying forces beyond its borders. With manpower "melting away," the Tsar ordered another massive round of conscription on December 12, 1812. To understand this act we must look at the resources of the French. Alexander's conscription was a precaution intended to provide additional resources to fight a defensive war. As contemporary observers made clear, a negotiated settlement was forecast, but what had Napoleon done? He had returned to France to raise another army. The Russians knew this and wanted to be able to negotiate from a position of strength. Even Gneisenau freely admits in his account that the Russians were not physically capable of maintaining themselves beyond the line of the Vistula River (see map 24).[21]

Napoleon had good reason to believe that he might return to Poland instead of several hundred miles further west. He could theoretically field more troops than the Russians, some 190,000. Napoleon's instructions to Murat, in command once the Emperor departed after crossing the Beresina, were to defend along the Niemen if possible.[22] However, the balance of troops in this total came from the Prussian and Austrian contingents—Murat's strength derived from the corps of two former enemies. If we subtract Yorck's 17,000 and Schwarzenberg's 25,000 troops from Napoleon's forces, the French Imperial forces for defense totaled 148,000, including those in garrisons. Napoleon, enraged by Yorck's action and underestimating the duplicity of the Austrians, further compounded his shortage of veteran manpower by directing Murat and his successor Prince Eugene to leave major fortresses garrisoned as the French gave up ground in their retreat west.[23] Instead of gaining power as they collapsed on their lines of communications, the French lost strength.

In mid-January, Yorck openly declared for the Russians and placed his forces at their disposal. Six weeks later, King Frederick William III fled to Silesia and secretly signed a treaty with the Russians at Kalisch. This treaty laid the cornerstone for the Sixth Coalition with the goal of liberating Germany based on the territorial configuration of Prussia in 1806. Great Britain rapidly concluded separate agreements subsidizing the Prussians and Russians and attempted to bring Swedish forces to Germany for use in a spring campaign. Austria remained neutral but began a massive mobilization, offering mediation to both sides.[24]

At this point we must digress before proceeding to the final military moves that ended this first phase of the campaign. Prussia had lost heavily during the Russian campaign. When her 17,000 are added to the Russian total, we find that the Allied field force was too weak to "liberate" Poland, much less Germany. Napoleon knew this. What he did not know about was Scharnhorst's secret mobilization system, which now went into effect. Mobilizing reserves and *Landwehr*, and with British

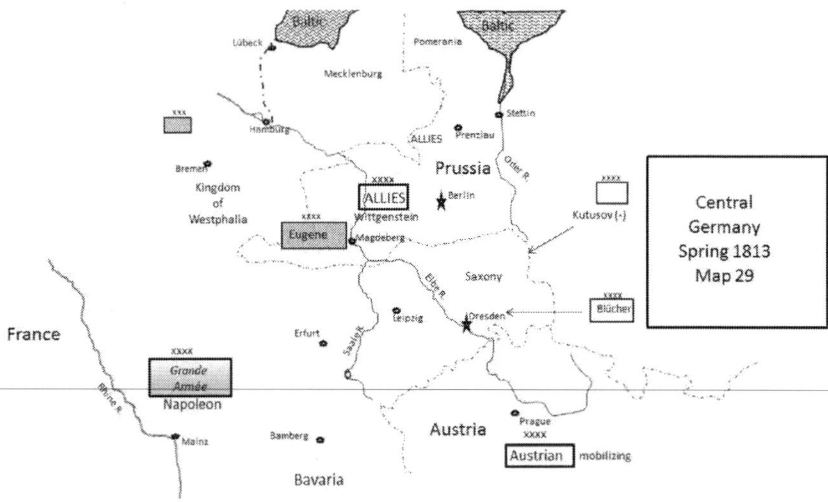

armaments, the Prussians fielded an army of 110,000 men prior to Napoleon's return to Saxony in April 1813. Of this total, the rawest recruits were diverted to help the Russians cover the fortresses occupied by the French. However, the best-trained, some 55,000, marched to join the main Russian army.[25]

The 1813 campaign up to this point had been one of operational pursuit. The frozen rivers of Germany provided Prince Eugene, now in command, no defensible barriers. He fought delaying actions as best he could as Germany exploded into revolt around his tired and demoralized troops. However, he was helped in no small measure by Kutuzov's lackluster pursuit. The Allied forces consisted of three field armies: one to the north under the Russian general Wittgenstein, one in Silesia under General Blücher, and the main Russian army under Kutuzov still in Poland (see map 29). Wittgenstein's army included his own corps and Yorck's Prussians—about 50,000 men. Blücher's army included a Russian corps under General Winzingerode. Other Prussian corps under Generals Friedrich von Bülow and Friedrich von Kleist were forming but not yet ready for operations.[26] The Allies planned the liberation of as much of Germany as possible, which by April meant as far west as the Elbe River. Here Kutuzov wanted to stop. In the meantime, Wittgenstein conducted a flank march along the Elbe to join Blücher who had captured Dresden. Eugene attacked Wittgenstein near Mockern, which threatened to upset the Allied plans to concentrate. However, after this indecisive combat Eugene withdrew and informed Napoleon of his intention to abandon the upper Elbe and retreat as far as the more defensible Saale River.[27]

We must take a moment to understand, briefly, how Napoleon reconstituted his new *Grande Armée*. First, he called up massive quantities of young conscripts, both from France and his allies, to make up his losses in veteran infantry. However, given the very common practice of "draft-dodging" in France, Napoleon took the drastic expedient of denuding his navy of its sailors and gunners, creating an entire army corps of *regiments de marine*. He remanned his artillery wholesale with experienced gunners from the fleet. Finally, his biggest weakness was his cavalry and with the loss of half of Spain that theater no longer had need for as many horsemen. Accordingly, he transferred regiments wholesale from there, both as entire units as well as cadres for new regiments destroyed in Russia. Many of these did not arrive until after the armistice in June. It was a feat of organizational genius, but its costs were heavy and only partially solved Napoleon's problems in recreating the sort of force he was used to wielding. Its chief advantage was its size, which was out of all proportion to what the Allies anticipated. Its chief disadvantages were its fragility, lack of cavalry, and overall lack of endurance—as we shall see, Napoleon's adolescent conscripts were unable to sustain the high rate of operations exhibited by his previous armies.[28]

The second phase of the spring campaign might properly be entitled the "battle phase." It was characterized by two fierce but indecisive engagements between the main armies of Napoleon and those of the Sixth Coalition. While Napoleon marshaled his forces at Erfurt for a resumption of the offensive, the Allies were thrown into a command crisis over Kutuzov's poor health. The Tsar appointed Wittgenstein as nominal commander in chief, but his authority only extended over his own corps and Winzingerode's. The Prussians, in a move to solidify unity of effort, generously placed their forces under Wittgenstein's control.[29] The corps of Miloradovitch and Tormassov, both senior to Wittgenstein, remained under the personal command of the Tsar. This confusing command structure was exacerbated by an operational dispute between the Prussians, led by Scharnhorst, acting as chief of the Prussian General Staff, and the Russian peace party formerly led by Kutuzov. Scharnhorst wanted to maintain the initiative and advance beyond the Elbe and disperse the French between the defiles of the Thuringian mountains and the Saale River before they gained strength—a lesson he had learned during the Jena campaign (see chapter 5). There were political imperatives as well: Austria was in contact and wanted the coalition to operate contiguous to her borders so that when she joined she would not have Napoleon between her and the main Allied forces. Obviously, the further west the Allies advanced the more time Austria would have to rearm and join them. Austria fully expected Napoleon to vigorously attack the Allies and wanted no part of any defeats until she was sufficiently rearmed.[30] This Austrian imperative wielded a significant influence over the remainder of the Allies' campaign and can be considered the dominant factor in the decision by the leaders of the Sixth Coalition to keep the bulk of their field forces in the southern part of the theater in southern Saxony and eventually Silesia.

Another consideration that weighed heavily in the short term involved the disposition of the Kingdom of Saxony, which had declared itself neutral despite its king's desire to throw in his lot with the Napoleon.[31] Kutuzov's death, Blucher's capture of Dresden and passage of the Elbe, and Wittgenstein's successful flank march persuaded Scharnhorst and the Prussians that a continuation of the offensive to the Saale was the best course. Baron Müffling, part of Scharnhorst's staff at the time, wrote, "We . . . placed our confidence in acting vigorously on the offensive, before Napoleon could unite with the Viceroy of Italy and fully develop his strength." The ambitious and aggressive Wittgenstein, backed by the Tsar, concurred and the decision was made to advance. They underestimated Napoleon's combined strength with Eugene by over 50,000, such were Napoleon's organizational skills in generating new armies.[32]

The Battle of Lützen resulted from Napoleon's resumed offensive. Napoleon's plan was to incur a battle or induce the Allies to retreat across the Elbe, either of which he was convinced would give the initiative and momentum back to him. To do this he planned to turn the Allied position by making "a movement exactly the opposite of the one I carried through during the Jena campaign" by marching on Leipzig then down the Elbe behind the Allies to Dresden.[33] The Russo-Prussian army surprised

Ney's isolated corps south of Lützen at the little village of Gross Görschen on May 2 (see map 30). Ney managed to hold on until Napoleon arrived with the bulk of his army of 110,000 men versus the 73,000 of the Allies. The Allied assumption that their veterans might provide the edge over Napoleon's conscripts proved incorrect and Napoleon forced his opponents to retreat. Each side lost approximately 20,000 casualties. Napoleon was unable to fully capitalize on his tactical success due to his paucity of cavalry. As if to underscore this deficiency Blücher led a furious cavalry counterattack that stopped the French pursuit cold and served as a postscript to this bloody battle. In reference to the fighting abilities of the Prussian, Napoleon allegedly muttered, "These animals have learned something." Unfortunately, Scharnhorst received a wound in the foot and died of blood poisoning some weeks later after turning over his duties as overall Prussian chief of staff to the capable Gneisenau.[34]

Napoleon had regained the initiative and the Allies retreated across the Elbe, but he had no operational reserves to press the issue, especially the "hell for leather" cavalry reserve of his heyday. The Allied plan remained, in Gneisenau's words, to "dispute every inch . . . to convince the Austrians that they were resolutely determined not to spare their powers nor . . . leave the deliverance of Germany entirely at the discretion of Austria."[35] The line of operations remained along the Austrian frontier and a defensive position was chosen in the foothills of the Bohemian mountains

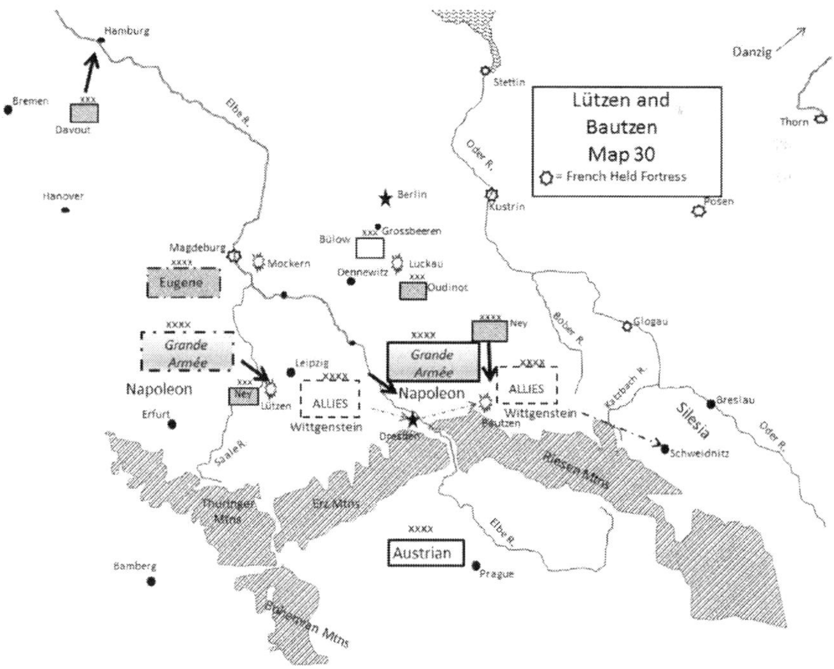

adjacent to the town of Bautzen from which to impede Napoleon's advance. Napoleon employed a classic operational *manoeuvre sur les derrières* against this position. His goal was to fix the Russo-Prussian army with his main army while Marshal Ney, whom he had dispatched to the north to threaten Berlin, moved south with two corps to arrive in the Allies' rear. Ney, or more probably his chief of staff Baron Jomini, directed all four corps (80,000 troops) against the Allied flank and rear at Bautzen. This fortunate accident was offset by Jomini's orders to the corps to march along the same route in column, thus limiting the coup de grâce to one narrow line of approach to the Allied vulnerable right flank commanded by the intelligent and resourceful Barclay de Tolly.[36]

The battle began on May 20 and all initially went well, with the Allies advancing deeper into Napoleon's trap. On the second day, Ney's four corps arrived with plenty of time to cut the Allied line of retreat and defeat their army in detail. But Ney further bungled the tactical execution by placing himself with the leading elements of the attack bogged down in the village of Preititz. Barclay informed Müffling of the impending disaster and weakness of his position and that officer convinced Gneisenau to persuade the Allied high command—Wittgenstein, Tsar Alexander, and King Frederick William III—to retreat. As the Allies retreated into Silesia they inflicted a series of sharp rebuffs to Napoleon's cavalry-deficient pursuit, particularly at Hainau on May 26 where one of Napoleon's infantry divisions was ambushed and effectively wiped out.[37]

In the aftermath, recriminations on the French side ran deep and Marshal Berthier preferred charges against Jomini (but not Ney). Jomini, in a huff, deserted to the Allies where the Tsar added him to his growing group of personal military advisors.[38] On the Allied side, Wittgenstein lost his job to Barclay de Tolly, whose conduct during the battle had earned him the laurels he long deserved. On the northern front, Napoleon peeled off another army under Marshal Oudinot to capture Berlin, a prerequisite to advancing to the Oder and relieving the French garrisons there. Oudinot's opponent was General Bülow who now had a sizable corps, well-armed with British equipment flowing through the Baltic and Hamburg. Oudinot and Bülow fought each other to a stalemate in the plains and forests southeast of Berlin, their last engagement occurring at Luckau on June 4. Thus, at the end of May, the Allies seemed to be in hopeless shape, their main forces had retreated into Silesia with Napoleon in pursuit and another force had narrowly averted the capture of Berlin. In the words of one observer the Allies were "absolutely in a cul de sac." Additionally, the coalition itself seemed in danger of breaking up over the issue of in which direction to retreat.[39]

However, Napoleon's new *Grande Armée* had no more offensive power left, being composed of many young recruits with little training and almost devoid of effective cavalry. Napoleon had robbed "Peter to pay Paul" by weakening his Spanish armies and had no more operational reserves on hand to finish the campaign. On June 4 at Pleischwitz the representatives of Napoleon and the Allies signed a cease-fire which later resulted in an armistice that was to last most of the summer. Why had Napoleon, on the verge of trapping the Allied armies in Silesia, settled for an operational pause? He explained it as, "lack of cavalry . . . and the hostile position of Austria." Napoleon's army was just as exhausted as the Allies and his experience in Russia had taught him the dangers of a protracted campaign beyond the capabilities of his army.[40] To fully understand Napoleon's problems we must address the other aspects of this campaign that had contributed to the French exhaustion.

The first of these, the problem of the fortresses Napoleon left along the Allied lines of communication, is disposed of rather quickly, but its bearing on Napoleon's problems is part of the entire mosaic. Napoleon's tasks for his fortresses were threefold: they were to tie down Allied forces; interrupt their lines of communication; and finally, because most of them were at major road and bridge centers, facilitate his own rapid movement when he advanced eastward. As far as the first task the reverse situation was true, Stewart remarks that the French garrisons were "nearly double the blockading force."[41]

Not only did the Allies mask these fortresses with minimal forces but also the troops used were often the least trained. These fortresses interfered relatively little with the Allied lines of communications during the critical period when the French withdrew for the reason that most of the rivers were frozen, rendering the bridges located at the fortresses irrelevant. By the time the rivers had thawed, the front was far to the west and the Allies had opened up new lines of support for their armies. Additionally, Napoleon's line of operations in the spring, in pursuit of the main Allied army, took him south from most of the fortresses. The example of Thom, which surrendered on April 16, provided other serendipitous benefits; captured artillery was immediately put to use in the Allied artillery park while the German partners of the French inside the fortress were paroled, many of them enlisting in the ranks of their former enemies (including a unit being formed by Clausewitz known as the Russo-German legion).[42]

The second set of operations involved the employment of numerous Cossacks, partisans, and friekorps by the Allies against Napoleon's communications. One Napoleonic historian credited these operations with shifting a balance of some 53,000 troops away from the French main forces for a cost of approximately 5,000 troops, mostly Cossacks. Their effect contributed measurably to Napoleon's operational problems. The first phase of these operations was an outgrowth of the way in which the Russians had been employing their Cossacks all along. Wittgenstein initiated these operations, which were a response to a request by emissaries of the Hanse cities and Eugene's withdrawal across the Elbe.[43]

The initial raid, by a mixed force of regular Russian cavalry, Cossacks, and expatriate Germans, succeeded beyond the Allies' wildest dreams. Its commander, Colonel Tettenborn, boldly advanced into Mecklenburg to secure the safety of Swedish Pomerania and gain the allegiance of the Duke of Mecklenburg-Schwerin to the Allied cause. However, the most stunning result was the withdrawal of the cowed French from Hamburg and the occupation of that strategic city by Tettenborn on March 18. Hamburg was later recaptured at the end of May, but while it remained in Allied hands, the British off-loaded tons of war material. These arms equipped the numerous Hanoverian, Mecklenburger, and other levies forming in this district under Count Wallmoden (formerly commander of the Hanoverian army).[44] The importance of the capture of Hamburg, the anchor of the French position on the lower Elbe, can be judged by Napoleon's response. He assigned possibly his most capable Marshal Davout the job of recapturing Hamburg and securing the lower Elbe. Davout, whom Napoleon could ill afford to do without, held down this assignment for the remainder of the war. Secondly, Napoleon held back a large percentage of his scarce cavalry, desperately needed for his campaign in Westphalia to protect that puppet kingdom from falling.[45] Following Wittgenstein's example, General Bülow detached Cossack forces under his control to cooperate with partisans and friekorps beyond the Elbe. These forces, led by General Doernberg (another Hanoverian) and the Russian general Czernicheff, proceeded to Luneberg where they destroyed the French division of General Joseph Morand on April 2–3. The Saxon troops of Morand's division, after the action, deserted en masse to the Russo-German legion of Czernicheff's command.[46] Wallmoden was appointed the commander of all these disparate forces in April and not long after Clausewitz became his chief of staff.

A final example of these raiding operations was the capture of Leipzig three days after the signing of the Armistice of Pleischwitz. Czernicheff, who had just destroyed a Westphalian column at Halberstadt, learned from his Cossacks that Leipzig, garrisoned by 5,000 French cavalry and infantry, contained numerous magazines and wounded. Czernicheff contacted the Russian General Woronzov, who was observing Magdeberg, and proposed a raid. Woronzov agreed to this plan and they were joined by the celebrated friekorps of Major von Lützow. In a brisk action on June 7, the combined forces of Woronzov, Lützow, and Czernicheff dispersed the French cavalry outside Leipzig and proceeded to occupy the city, only to learn that an armistice had been concluded three days earlier.[47] Woronzov's raid reveals just how tenuous Napoleon's lines of communication through Saxony were. If we wonder as to Napoleon's weakness in cavalry, the above operations show that a good portion of it was employed guarding the lower Elbe and the French rear, and not very effectively either.

In review, one sees in this campaign the standard Napoleonic format: a search for decisive battle along a single line of operations, in this case north of the Bohemian mountains. However, a cursory examination reveals a more complicated picture. The operations of the Allied forces during this phase of the campaign were distributed, displaying a level of sophistication not previously seen in central Europe. The most compelling proofs of operational distribution were these raids, harassment, and the taking and holding of key cities by the forces of Tettenborn and others. That these operations were the outgrowth of the Russian experience, and largely opportunistic, does nothing to diminish their operational effect. Napoleon now had another front that he had to honor—in his strategic and operational rear. The events leading to the armistice following Pleischwitz provide the evidence that both sides were exhausted and meet the criteria of modern operational campaigns being exhaustive in the Delbrück sense (see chapter 1).[48]

The Allies did not rest on the serendipitous results of the first raids, but continued to press the French and bring a new level of organization to these forces by appointing Wallmoden to command and synchronize their operations. Finally, the commitment of substantial French forces for security—including much of their cavalry—can be considered evidence of operational art because it denied Napoleon the tool he needed to engage the Allies throughout the depth of the battlefield following his tactical victories at Lützen and Bautzen. Also, the Allies employed continuous logistics and distributed deployment, which can be said to have "arrived." Witness the mountains of British equipment the French captured when they retook Hamburg. This equipment gives only some indication of the far larger amounts that made it into the hands of the Hanoverian, Mecklenburger, and Prussian levies to say nothing of the rearmament of the Cossacks with better equipment.[49] The British not only armed the Germans, but they did so on the Germans' home ground, thus integrating the processes of mobilization and deployment by avoiding the classical practice of marching unarmed levies to a depot or collection point. Napoleon performed the same extraordinary task, arming and training his young conscripts on the march—and without the advantages conferred on the Allies by command of the sea to transport supplies.[50] The battles of Lützen and Bautzen show that the Allied armies had reached new levels of operational durability. This durability was admittedly immature; they were exhausted after six weeks of fighting—but so were the French. However, when we factor in that the Russians had been fighting for almost a year without pause, we realize that the Russians may have been more durable than their Prussian ally. Imagine the coalition army of 1805 holding together after a Bautzen or a Lützen? Thus, the Allied coalition had become more *politically* durable.

Two areas that superficially appear less mature than the others discussed are operational vision and instantaneous command and control (C2). Instantaneous C2 is the more problematic of the two. The only medium of "instant" communications was that of the shared vision of the participants. All wanted to fight the French, hurt them as much as possible, and ostensibly liberate Germany. This led to an environment that favored commanders willing to make independent decisions. Such rapid decision-making in response to opportunities yielded the most fruit in the deep operations along the Elbe and in Napoleon's rear. As for operational vision, the Allies were hurt by the deaths of Kutuzov and Scharnhorst, particularly Scharnhorst if we are to believe Müffling, Clausewitz, and Gneisenau. But we catch a glimpse of Scharnhorst's operational vision as we see the product of his creativity in action—the mobilization and performance of the Prussian army.[51]

The survival and will to fight of the coalition army after Lützen and Bautzen and the shared vision remaining in force after Scharnhorst died provide further evidence. Perhaps the true measure of operational vision is the ability to pass it along to one's peers and subordinates, rather than merely to rely on electronic communication. We must also recognize that the coalition itself had improved and was improving.[52] The effort the Sixth Coalition expended in maintaining its unity was unique for the Napoleonic Wars. Blücher's willing subordination to Wittgenstein is but one example. Even the choice of the line of operation, south away from the threatened capital of Berlin, was a departure from the classical strategy of the past. It would be like Kutuzov retreating toward Kiev instead of Moscow after the battle of Borodino. The signing of the armistice in June 1813, too, has a modern flavor. The Allies could well have kept fighting, which might have kept Austria out of the war, or done as Austria did after Wagram and negotiated an unfavorable peace. Instead they chose the best solution, an operational pause in the form of a cease-fire and then armistice. This represented a collective vision; coalition unity for the sake of ideals—liberation for Germany and the restoration of the balance of power. This vision provided fertile soil for operational commanders, but the harvest was yet to come.

One must also give "the God of War" his due.[53] Napoleon was learning new ways of war, despite his ongoing search for the decisive battle that ended the campaign. The battle of Bautzen provides the best evidence of Napoleon fighting in a distributed fashion and further evidence of his continued mastery of operational maneuver. Napoleon, like Robert E. Lee at Chancellorsville, assigned himself the job of fixing the Allies, while Ney delivered the coup de grâce from the flank with another entire army. Napoleon had been effectively fighting as an army group commander since Russia and this campaign shows that his sophistication at maneuvering armies was increasing despite the degradation of his tools. Finally, Napoleon recognized that he must respond to the operational threat to his rear and flanks, another lesson from Russia. The Emperor allotted his precious cavalry and arguably his most capable marshal to these tasks. However, Napoleon's own distributed efforts against the Allied rear failed primarily because of the sophistication of the Allies' response. His strategy of retaining fortresses evoked a nonclassical response from the Allies—they simply masked these fortresses. Nevertheless, Napoleon, too, was evolving, and fighting in a more distributed manner.

THE OPERATIONAL PAUSE—TRACHENBERG AND REICHENBACH

On October 9, 1813, Napoleon arrived with 150,000 troops opposite the town of Duben on the Mulde River. He had made the decision two days earlier to march west from his position at Dresden in order to intercept two coalition armies that had united and moved into his rear south of the Elbe River; one under Field Marshal Blücher and the other under the former marshal Bernadotte. Napoleon's advance guard had engaged the rear guard of Blücher's army (Sacken's corps) for most of the day and as these combats died down and darkness closed in, Napoleon realized that the Allies had refused battle. Where had the Allies gone? More importantly why had two Allied armies, both recently victorious in combat against Napoleon's flank armies and with approximate numerical parity in numbers of troops, refused to give Napoleon battle?[54]

Historical hindsight provides a clue. One week later, these same Allied armies, particularly the aggressive Blücher's Army of Silesia, effectively united with an even larger Allied force, the army of Bohemia (180,000 troops), advancing from the south in the sprawling Saxon countryside around the city of Leipzig. During a bloody battle of attrition from October 14–19, these forces, in Europe's largest battle to date, defeated Napoleon and forced him to withdraw completely from Germany west of the Rhine. This result was no accident, and had in fact been outlined in an operational plan for the campaign developed earlier that July in Silesia. The Allies had refused battle in order to maneuver as planned to gain more favorable conditions for an engagement in the future.

The situation after the armistice, it will be remembered, was one of acute exhaustion on both sides. Gneisenau freely admitted as much, stating that "Austria had given Russians and Prussians to understand that such a period of time was necessary to complete her armaments." He additionally listed eight objectives that he felt needed to be accomplished in order to renew the combat with Napoleon: reinforce the Russians, principally through a new army forming in Poland under Count Bennigsen; complete the manning of the Prussian line regiments; obtain arms and ammunition from Great Britain and Austria; prepare "accoutrements" (uniforms); arm, form, and discipline the landwehr; provision and repair fortresses; establish bridgeheads on the Oder river (the main crossing sites were in French hands); and procure and collect provisions.[55]

The advent of Austria as an active belligerent against Napoleon was by far the most important development during the armistice. Ironically, this goal, and many Gneisenau listed, was shared by Austria's foreign minister Clemens von Metternich who commented upon the armistice as follows: "An Armistice will be the greatest of blessings . . . it will give us an opportunity to get to know each other, to concert military measures with the Allies and to bring reinforcements to the most threatened points."[56] From this quotation it appears that Austria's participation was a forgone conclusion. However, as Charles Stewart observed, "It is difficult to give an adequate idea of the anxiety that prevailed [on the part of Great Britain, Russia, and Prussia] . . . with respect to the decision of Austria."[57] Metternich's master, Francis I, was not as committed to war with Napoleon as his minister and hoped that the armistice might give Austrian diplomacy the opportunity to negotiate a lasting peace. The terms to establish a peace comprised: the return of Illyria (the Dalmatian coast) to Austria, the territorial aggrandizement of Prussia via the dissolution of the Grand Duchy of Warsaw, the break-up of the Confederation of the Rhine, and the reestablishment of France's eastern boundary along the Rhine. The acceptance of these terms by Napoleon would have guaranteed Austrian neutrality and certainly freed Germany geographically from French influence. On June 26, 1813, Metternich had presented these terms to Napoleon at Dresden and was rudely rebuffed; Napoleon would accept the Illyrian concession only. Metternich departed Dresden convinced that peace with Napoleon was not possible.[58]

From Dresden Metternich proceeded to Allied headquarters in Reichenbach, Silesia, where he committed Austria to join the Sixth Coalition should Napoleon remain intransigent. Since Napoleon had already made clear his intent, the Treaty of Reichenbach was in effect Austria's declaration of war. Now the Allies proceeded to plan in earnest for a resumption of hostilities with Austria as a cobelligerent (the armistice had been extended into August). It was at this moment that Wellington's summer offensive in Spain climaxed. He engaged King Joseph's army at Vitoria with his Anglo-Spanish-Portuguese army and routed it. Although French troop losses were not that large, the material and moral consequences of Wellington's victory had the strategic effect of expelling the French once and for all from Spain. It could not have come at a worse moment for Napoleon and galvanized the Sixth Coalition, increasing Emperor Francis's conviction that he must throw in his lot against Napoleon. However, no single treaty of grand alliance linked all the players together.[59]

Against this backdrop the Russian, Prussian, and Swedish high commands met at Trachenberg in Silesia to decide the command structure and work out a plan of operations. Barclay, still the nominal commander for the Russo-Prussian army, had been working off a series of plans that resembled some of the elements of Russian plan of the year before. Not surprisingly, he built on the idea that as Napoleon advanced against one army, the other—a Russo-Prussian force—would attack his flank. Underestimating Napoleon's ability to organize, Barclay predicted a clear superiority over the French.[60] As it turned out, the accession of Sweden to the coalition resulted in three armies. A "galaxy" of diplomats and leaders of the Sixth Coalition descended upon the castle of Trachenberg in Silesia on July 9–12 to discuss acting "in concert in the distribution of their forces" and create "a . . . general plan of operations."[61] Attendees included the Tsar, King of Prussia, and Crown Prince of Sweden (Bernadotte) and their personal military advisors. Gneisenau, Blücher, Schwarzenberg, his chief of staff Josef Radetzky, and the expatriate Jean Moreau, who advised the Tsar, were not there—but their input was. The names of the generals who were actually there are not as well known: Lowenhielm (Swedish), Toll and Volkonsky (Russian), and Knesebeck (Prussian).[62] According to Baron Müffling, the conference resulted from the Tsar's desire to resolve command issues revolving around Bernadotte.[63]

Bernadotte was still awaiting Prussian and Russian ratification of the treaty he had signed in April that guaranteed Sweden compensation in Norway at the expense of Napoleon's Danish ally. He had not been notified of the cease-fire and subsequent armistice and had threatened to withdraw from the coalition. Bernadotte's terms now included command of one of the principal Allied armies. In addition to the approximately 110,000 Swedes, Prussians, Russians, British, and North German troops already under his command, Bernadotte demanded command over Blücher's Army of Silesia.[64] Bernadotte's "demands were too great, and could not be conceded by the sovereigns. They wished, however, to see him return satisfied from Trachenberg . . . and admitted that *circumstances might* [italics added] render it necessary for him also to take the command of . . . the Silesian army."[65] This compromise could have had a negative impact during the upcoming operations because it created an environment for conflict with Blücher over combined command of the armies. However, Blücher and the dynamic Gneisenau conducted upcoming operations with little reference to Bernadotte; part of their success would be due to that factor as well as the fact that Bernadotte's chief subordinate, General von Bülow, often initiated engagements that his master would have avoided had he commanded von Bülow's corps![66]

Blücher's Silesian army was to be the smallest of the three because nearly 100,000 troops, Russians and Prussians, were to accompany the Tsar of Russia and the King of Prussia to join the Austrian army of Bohemia—this latter conglomeration eventually came to a strength of over 220,000 men and can rightly be considered an army group.[67] Tsar Alexander hoped for the appointment of Archduke Charles as the commander of the main Austrian field army, who would thus also serve as commander of the combined army formed when the Russians and Prussians joined forces with the Austrians in Bohemia.[68] In May, Francis and Metternich appointed Prince Schwarzenberg as the Austrian commander in chief. The Tsar subsequently nominated himself as supreme commander of the combined armies in Bohemia. Metternich countered with the argument that the country with the preponderance of force should command the main army. On August 6, the issue came to a head. Metternich threatened to maintain Austrian neutrality should Alexander replace Schwarzenberg as supreme commander. The Tsar reluctantly acquiesced in this decision.[69]

The problem with this arrangement was that Schwarzenberg assumed command at three levels with only one undermanned staff led by Radetzky, the Austrian chief of staff (see figure 4). The first level was that of the Austrian field army—nearly the same

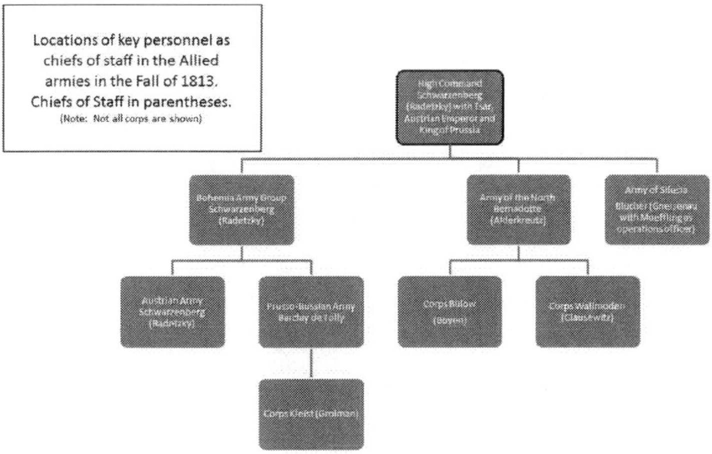

FIGURE 4

size as Charles's army at Wagram. The next level up—the Army of Bohemia—added the Russo-Prussian army commanded by Barclay. This force was effectively an army group, but no additional staff was provided to control it. This was a problem the Allies never resolved; as with the Swedish issue, command of the Army of Bohemia raised its ugly head repeatedly during the upcoming campaign. Finally, Schwarzenberg and Radetzky were nominally responsible for the operational direction of Blücher and Bernadotte's armies, a task far beyond their means or inclination and that they wisely left to the expedient of decentralized execution, much as the Russians had done in 1812. The choice of Schwarzenberg proved a good one despite critics of his generalship. It was Schwarzenberg who had recommended attaching Russian and Prussian corps to the Austrian army to further improve unity of effort.[70] Historian Gordon Craig's evaluation of Schwarzenberg summarizes his suitability for a multinational command:

> The new supreme commander's talents were . . . more diplomatic than strictly military, and it was probably a good thing that this was so. Like Dwight D. Eisenhower in another great coalition a hundred and thirty years later, his great gift was his ability, by patience and the arts of ingratiation, to hold together a military alliance which before Napoleon was finally defeated comprised fourteen members, and to persuade the quarreling monarchs and their field commanders to pay more than lip service to the alliance's . . . plan.[71]

The plan agreed to at Trachenberg envisioned three main armies. The two larger armies, under Bernadotte (90,000) and Schwarzenberg (220,000), would threaten

Napoleon's flanks from the north and south. Blücher's smaller Silesian army (50,000) would face Napoleon to the east but had specific instructions to "avoid committing itself except in the case of an extremely favorable situation." Any two armies not engaged by the French main effort were to attack the French flank, rear, and lines of communications.[72]

The Convention of Trachenberg (see appendix) was forwarded to Schwarzenberg and Radetzky after the conference in July. Radetzky was another reason that decision to give the supreme command to the Austrians proved fortuitous. Up to this apoint, Radetzky had been a quiet but effective force for reform in the Austrian army and led Austria's mobilization and planning once the extent of Napoleon's disaster in Russia became known. When Radetzky saw his initial operations plan, which assumed a successful advance to the Rhine, obviated due to the sledgehammer blows of Napoleon during the spring of 1813, he began work on a new plan to defeat Napoleon.[73]

Radetzky explained his plan to the British envoy Sir Robert Wilson as a "system of defense combined with offensive operations on a small scale over a general offensive movement which might win much, but also might lose all."[74] Radetzky modified the Trachenberg Convention in the following important ways. First, the plan he received predicted an immediate offensive by Napoleon toward Bohemia. It therefore directed Bernadotte to energetically advance south and combine with Blücher to form a second army in Napoleon's rear of about 120,000 men. This was done to make Bernadotte happy but Radetzky wanted three independent field armies operating against Napoleon until he could be engaged with a clear superiority (presumably by the Bohemian army group). Secondly, no allied army would go on the offensive against Napoleon initially. Neither of the two smaller armies was to engage him, but rather retreat energetically, causing French forces to make forced marches in pursuit. This meant that the Allies would react to Napoleon's movements as they employed a strategy of exhaustion, making his forces march and countermarch until battle could be accepted on favorable terms. It ceded to Napoleon his favorite principle of war—*initiative*. However, the Austrians intended to use Napoleon's strength against him, to have him spend his initiative and energy on the offensive, thereby reaching, as he had in Russia, a culminating point of victory far from France, resources exhausted, and surrounded by hostile armies. Radetzky and Schwarzenberg forwarded their changes from their headquarters to Reichenbach on July 19 and these were adopted verbally, although never finalized in written form. Thus the Trachenberg Convention became, properly, the Reichenbach Plan.[75] The Sixth Coalition's leaders created military history's first modern operations order for a multinational army group. There were other inputs that affected the plan's operational execution. General Jean Moreau, one of the Tsar's many advisors, for example, advised: "Expect a defeat whenever the Emperor attacks in person. Attack and fight his lieutenants whenever you can. Once they are beaten, assemble all your forces against Napoleon and give him no respite."[76] Jomini, now one of the Tsar's aides, cautioned against underestimating Napoleon—stating that he remained "the ablest of men."[77]

Finally, the equipping and training of the Allied armies during the armistice, including a 10-week rest for the veterans, yielded an immense force for the coming campaign. The Army of Bohemia passed in review on August 19 outside Prague for the monarchs of Austria, Prussia, and Russia. The British observers, whose government had largely paid for much of what was on display, noted that the funds had been well spent and were satisfied with the state of the soldiers and equipment.[78] In central Europe alone the coalition deployed over 570,000 soldiers against approximately 410,000 troops of the reconstituted *Grande Armée*.[79] This disparity is even greater given the more than 50,000 veteran French troops cut off in fortresses by lesser numbers of second-line troops. Napoleon's *Grande Armée* of August 1813 was a more potent force than the one he had marched and fought to exhaustion in the spring. His Imperial Guard and cavalry had been reconstituted by extraordinary efforts, especially by the expedient of denuding the armies in Spain. Napoleon's artillery was also numerous and excellently served (often by former sailors)—but not nearly as mobile. However, his army remained overwhelmingly young—two-thirds of his troops were between 18 and 20 years of age. About 90,000 of these troops would be on the sick lists before hostilities began.[80] The many foreign troops, especially Germans, exacerbated these factors because they composed a significant number of Napoleon's veterans. Napoleon compounded his problems by apportioning the majority of these German troops to his flank forces, just as he had in Russia. These forces predominantly opposed fellow Germans instead of the more ethnically diverse force of the Bohemian army. Finally, Napoleon was forced by circumstance to employ the best of his remaining independent commanders elsewhere: Eugene in Italy, Soult and Suchet in Spain, and Davout holding the lower Elbe. The only exception was Marshal St. Cyr, who kept guard at Dresden.

One more command arrangement must be discussed before proceeding to the operations—the Prussian General Staff system. Although the Prussians commanded the numerically smallest of the field armies, they wielded a significant operational influence across the coalition. Gneisenau took what might be seen as a disadvantage, the breakup and distribution of the four Prussian army corps among the three armies, and turned it to his advantage. He did this by assigning competent Quartermaster General-trained staff officers as chiefs of staff to all of the main corps and even some of the non-Prussian ones. For example, Clausewitz, still in a Russian uniform, served as chief of staff to Count Wallmoden, commanding the corps observing Davout in Hamburg.[81] Hermann von Boyen, a major reformer of 1807, was assigned to the critical corps of General Bülow serving under Bernadotte. In the same manner, Karl Grolman, also a reformer and former Prussian minister of war, was assigned to the corps of General von Kleist serving with the Bohemian army (see figure 4). Both Grolman and Boyen had been protégés of Scharnhorst and part of his intimate circle. Gneisenau instructed these men to coordinate, as much as possible, their actions with his in Silesia. If problems arose the chief of staff "had special avenues open to him. He communicated any complaints or doubts directly to the chief of the General Staff himself."[82] They overcame the difficulty of communication over the vast theater that spread around Napoleon's salient in Saxony by this shared understanding, which was transmitted in the Reichenbach Plan and managed in execution by Gneisenau.

Meanwhile, the sham "peace talks" in Prague collapsed on August 10—despite Napoleon's agreement to most of the Allies' terms. Three days later, Austria declared war on France.[83] By this time, the Allies had a second operational echelon approaching under Bennigsen, the Army of Poland with 60,000 recently raised Russian troops. Napoleon had to act before these arrived in the theater. Instead of advancing directly on Bohemia as the Allies expected, however, he had Marshal Oudinot advance on Berlin with 90,000 troops while Napoleon remained in two operational lines around Dresden and just to the northwest of Dresden with 250,000 men, including his Guard and cavalry reserves.[84] Napoleon's apparent inaction caused the Allies to reconsider their plan. They had not allowed for an inert defense by their opponent, but anticipated a move on his part in order to react with the armies not opposed to him. A "general offensive" movement, contrary to Schwarzenberg and Radetzky's desires was agreed to by a council of war. Schwarzenberg had planned the logistics for the Austrian army to support an eventual advance on Leipzig, and now that an offensive was to be conducted, he naturally recommended Leipzig as the objective. Orders were sent and the Anny of Bohemia began to advance. The Tsar, advised by Moreau, now interfered with the arrangements of the nominal commander in chief. Alexander and Moreau felt a move closer to Blücher in Silesia warranted, and indeed that was where Napoleon had gone in response to an advance by the Prussian firebrand. The Tsar's view prevailed, despite the opposition of Schwarzenberg, and Dresden became the new Allied objective. Schwarzenberg had considered moving on Dresden as well but had wanted to take advantage of his logistics preparations and wheel on the city after advancing through the Bohemian mountains. Metternich, responding to Schwarzenberg's consternation over these events, wrote, "the most sincere understanding between us and our allies is so important that we cannot offer too great a sacrifice" (see map 31).[85]

Logistics support, set up for an advance on Leipzig, soon broke down in the advance to Dresden. The effects of countermarching and the wet, rainy weather further fatigued and slowed the advance of the Allies. The lead elements of the Army of Bohemia arrived, cold, tired, wet, and hungry, south of Dresden on August 25. Napoleon was not yet there. Instead of attacking while Napoleon was still absent,

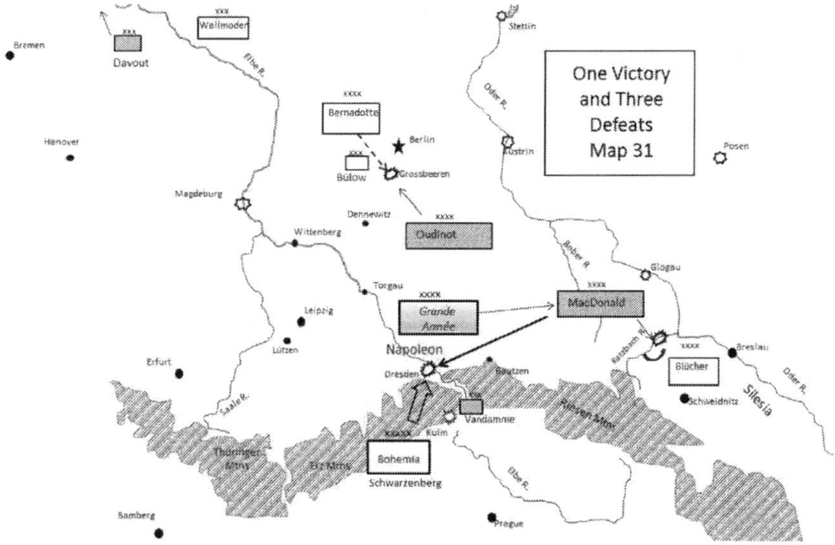

another war council was held by the "military college" accompanying the army.[86] Schwarzenberg and Jomini supported the Tsar's recommendation for an immediate attack, but Moreau and Toll advised against it.[87]

The attack was put off until the next day (August 26). Marshal Gouvion St. Cyr faced the Allies at Dresden. He had earned his marshal's baton in Russia at Polotsk fighting just the type of battle the Allies now contemplated. Allied skirmishers had already found Dresden's walled houses and gardens fortified for the impending attack. As the initial combat into these fortified suburbs began, Napoleon dramatically arrived. With Napoleon's arrival the mood at headquarters rapidly changed and Moreau advised Alexander to withdraw. The Prussian King uncharacteristically asserted himself and urged that the attack continue. While the supreme command bickered, the assault went on in response to orders already issued. The result was a defeat for the coalition as Napoleon crushed the wings of the Bohemian army group. Already half-beaten, the Allies compounded their mistake and fought a second day until forced to withdraw on August 27. Their losses were significant: 38,000 Austrians, Prussians, and Russians were casualties, including many prisoners, versus 10,000 French casualties.[88]

Dresden was the exception that proved the rule. The Reichenbach plan had not intended an offensive battle against Napoleon and his main force by one army—particularly against a defensive position like Dresden. The Allied leaders, partially as a result of the ponderous command process of the Bohemian army, diverted in spirit from the agreed plan, and fought Napoleon on his terms. Moreau, whose legs were shot off by a cannonball, tried to put a bold face on the defeat as he wrote to his wife with his dying hand, "That scoundrel Bonaparte is always fortunate. . . . Though the army has made a retrograde movement, it is . . . from want of ensemble, and in order to get nearer General Blücher."[89] The British observer Wilson was less charitable in his assessment; he called the battle "an ill-advised enterprise executed with great vigour."[90] Dresden validated Napoleon's improvements to his army during the armistice. His Young Guard had resolutely defended the city on August 26 and his cavalry and horse artillery had been critical in the counteroffensive that forced the Allied withdrawal the following day. However, Marshal Marmont had earlier (August 16) expressed to Napoleon his concern about fighting on such a widely extended front with the prophetic words, "I greatly fear lest on the day on which Your Majesty gains a great victory, and believes you have won a decisive battle, you may learn you have lost two."[91] These concerns literally came true.

The Allies' operational scheme now paid handsome dividends on other fronts. Recall that Marshal Oudinot, opposed to Bernadotte, had been assigned the mission of taking Berlin. He began his advance on August 18. He initially gained some minor successes but was forced to divide his army into three corps-sized columns as he advanced in the rain through the heavily wooded terrain south of Berlin. At the same time, Davout advanced in support from the west, including Girard's division from Magdeburg. Bernadotte's reaction was typical of a sovereign whose chief concern was the preservation of his army—he built additional bridges over the Spree River to facilitate a withdrawal to the north with the intention of abandoning Berlin.[92]

Bülow, the commander of Bernadotte's Prussian corps, refused to abandon Berlin, possibly as a result of previous coordination with Gneisenau.[93] Oudinot's separated columns offered Bülow an opportunity to attack the French while they were dispersed with wooded country to their rear. Oudinot's right-hand column under General Bertrand successfully repulsed an attack by General Tauenzien's Prussian landwehr at Junsdorf on August 22. Bertrand pursued on August 23 and was in turn repulsed by Tauenzien at Blakenfeld. This defeat prevented Bertrand from coming to General Reynier's aid when Bülow attacked his Saxon Corps, Oudinot's central column, in a driving rain at Grossbeeren on August 23. Reynier withdrew before Oudinot, personally leading the left-hand column, could come over in support. Oudinot's entire army was forced to fall back as a result. The domino effect also extended to Davout, now exposed by Oudinot's retreat, and he withdrew to Hamburg. Girard was most unlucky in these series of setbacks; he was caught in an isolated position and savaged by the ubiquitous Czernicheff and his Cossacks on August 27, losing almost half of his division in the process.[94]

Likewise, in Silesia, Blücher's adherence to the Reichenbach plan led to victory on a larger scale. As in the north, difficulties between the commander and his subordinates caused problems. The situation in Blücher's army was the reverse of that in

Bernadotte's—it was his subordinate, the French émigré General Louis Langeron, who was the more cautious. Blücher was under the nominal command and control of Barclay de Tolly, who had been delegated the responsibility of coordinating the operational movements of the Army of Silesia with respect to the Army of Bohemia by Schwarzenberg. Barclay and Blücher had met on August 11 at Reichenbach to ensure the proper coordination and understanding prior to the Russian's departure for Bohemia. Blücher's instructions were not to become decisively engaged with the force facing him. He proposed a more aggressive role for his army that involved attacking the French if Napoleon was confirmed not present *and if* the French had not attacked first. This course of action was approved by Barclay. However, Barclay neglected to inform Langeron of this change. As a result, Blücher's energetic advance on August 20 against the French was robbed of success when Langeron refused to cooperate in an effort to cut off an isolated corps under Ney near the Bober River.[95]

Meanwhile, Napoleon arrived on August 21 with considerable reserves to confront Blücher. Blücher, greatly outnumbered, retreated according to plan to a previously prepared defensive position on the Deichsel stream. Advancing beyond this position, Langeron undermined Blücher's plans yet again by falling back as soon as he was attacked on August 22. This precipitate retreat undermined the integrity of Blücher's position and he again withdrew to a position behind the Katzbach River. Blücher denied Napoleon a major battle and French casualties roughly equaled those of the Allies—attrition this manner favored the Allies. Almost 5,000 Allied troops were lost as a result of these misdirected actions.[96]

When Schwarzenberg's lumbering advance on Dresden pulled Napoleon and his reserves away to assist St. Cyr, Napoleon left Marshal Macdonald in command of his own corps and three others (including Sebastiani's cavalry corps) with strict orders not to advance beyond the Katzbach. Heedless of these instructions, Macdonald pursued Blücher beyond the Katzbach on August 27.[97] It was at this time that Blücher noted a lack of aggressiveness in the French approach and pursuit, correctly deducing that Napoleon had departed. He and Gneisenau resumed the offensive in accordance with their agreement with Barclay. The result was a meeting engagement along the Katzbach, which a driving rain had swollen into a raging torrent. Blücher waited until about half of Macdonald's army was across before attacking in force on the plateau above the stream. The portion of Macdonald's army on the Prussian side of the river was totally defeated and thrown into the river. Many French drowned as they attempted to recross and Macdonald lost over 15,000 men and many cannon. Worse, the army disintegrated in the cavalry pursuit that followed, although many of these men later returned to the colors. Gneisenau's comment on the operations of the Silesian army illuminates the perceptions of those present: "In the course of eight days [Blücher] had fought [Macdonald] in eight bloody encounters, not to mention trifling affairs; beat him completely in a pitched battle and directly afterwards made three serious attacks upon him."[98]

The results of the Katzbach and Grossbeeren clashes validated the Reichenbach Plan, despite the serious setback at Dresden. Local disunity in both of the smaller armies had threatened failure: Bülow's conflict with Bernadotte and Langeron's insubordination vis-à-vis Blücher. Nevertheless, operational unity due to general adherence to the operational framework yielded two significant victories that effectively negated the results of Dresden, just as Marmont had predicted. More victories followed. Despite the pounding at Dresden, the Army of Bohemia conducted a fighting withdrawal south through the forests of the Bohemian Mountains. Partly due to tenacity, and partly due to luck, a victory was obtained on this front, too. After Dresden, Schwarzenberg complained bitterly to Metternich and wrote that either the Russo-Prussian Reserve Corps be placed "under my immediate orders, or someone else be entrusted with the command." While Schwarzenberg vented to Metternich, he and Radetzky resolved to halt the French pursuit by General Vandamme that threatened to cut off major portions of the Allied army group. On August 29, at Preisten (just south of Kulm), Schwarzenberg turned and fought, sacrificing the Russian Guards in a vicious counterattack that halted Vandamme for the moment. The next day, Vandamme renewed his assault, but Kleist's Prussian Corps, which had been cut off by Vandamme, unexpectedly debouched on the Frenchman's rear at Kulm. Vandamme was taken prisoner and more than half his large corps annihilated or captured—crushed between Allied forces advancing both up and down the narrow valley.[99]

On the northern front, Bernadotte, encouraged by Bülow's victory, cautiously advanced toward the Elbe on Napoleon's northern flank. Napoleon preferred to face his former marshal personally, but had to rescue Macdonald in Silesia. Accordingly, he replaced Oudinot with the aggressive Marshal Ney who immediately resumed the offensive toward Berlin. On September 6, Ney stumbled into a trap that Bülow laid for him north of the Elbe at Dennewitz. The fighting followed a characteristic pattern: energetic Prussian and Russian attacks with Bernadotte holding his precious Swedes in reserve. Nonetheless, the Allies won the battle, shattering Ney's army in the process.[100] Because of the fragility of his young conscripts, the durability of Napoleon's corps and armies proved far less than that of earlier campaigns and at times it seemed that only he could rally the troops and put his humpty-dumpty armies together again.[101]

To appreciate Napoleon's operational problem one only need follow his movements in September. Early that month he had moved from Dresden east to Macdonald's army to halt, and hopefully defeat, the advancing Blücher. That general reacted by conducting a fighting withdrawal, correctly assessing that Napoleon and his operational reserves were once again with Macdonald. Napoleon then proceeded again to

Dresden in response to St. Cyr's call for help against the Army of Bohemia, which was again advancing after having recovered and reorganized. The Bohemian army had effectively split into two separate maneuver columns; one under Schwarzenberg and one under Alexander and his advisors. However, they had learned from Dresden and they separately withdrew once Napoleon's presence was known. Napoleon considered pursing Alexander's Russo-Prussian column, which was just out of supporting range from the Austrians, but he was again called away after the disaster at Dennewitz. Before he could deal with Bernadotte, he learned that Schwarzenberg was advancing again, this time to Pirna—and rushed there to contain the threat. He stabilized that situation only to learn of a renewed advance by Bernadotte to the Elbe. While Napoleon hurried north to deal with this problem he was further diverted to the east to again deal with Blücher. On September 22, he repulsed Blücher's advanced guard in the vicinity of Bautzen (see map 32). Blücher again withdrew in response to Napoleon's presence.[102]

In this manner the Allies prevented Napoleon from regaining the initiative. Napoleon's comings and goings included a corresponding movement of his reserves. These marches and countermarches had the impact of another defeat by detracting from the strength of the *Grande Armée*. Thousands of Napoleon's young conscripts dropped out of ranks, including in the Guard. Hunger also became a serious problem as the rapidly shifting moves outstripped Napoleon's logistical arrangements. The Allies

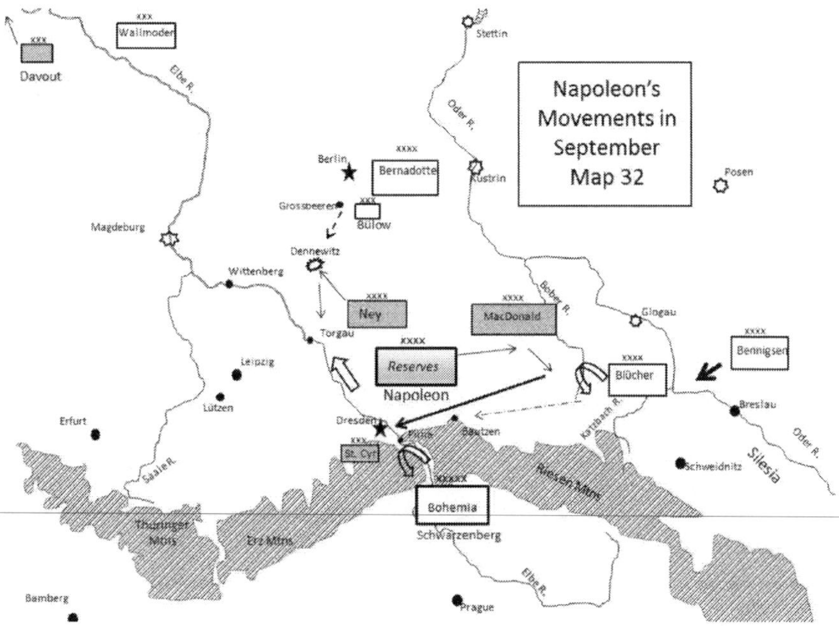

were well informed of Napoleon's deteriorating situation. The ubiquitous Allied cavalry and raiding *Freikorps* obtained vital intelligence on the *Grande Armée*'s dispositions, intentions, and morale. Wilson referred to this lucrative information source as "an infinity of intercepted official and private letters."[103]

In previous campaigns Napoleon had usually had the advantage in operational intelligence, but not in 1813. As in the spring phase, Allied raiders wrought havoc on Napoleon's communications, causing him to detach major formations to deal with them. A few examples suffice to illustrate that this phase of the campaign included many, if not more, of the deep operations seen earlier that year. Blücher "harassed the rear of the French Army" using the Cossacks attached to the Russian corps in his army. One enterprising Cossack leader captured an entire French infantry battalion at Wurschen on September 2. These Cossacks also provided Blücher a wealth of intelligence including the condition and intentions of the Poniatowski's Polish corps.[104] Both Stewart and Gneisenau emphasize the importance of these operations on Bernadotte's front. They included Cossacks, raiding *Freikorps*, and partisan activity. One such force under General Thielmann bagged 1,300 prisoners at Weissenfels, west of the Saale River. The booty also included dispatches relating "the most doleful details of the French Army."[105] Napoleon's communications were so disrupted by this activity that he detached a cavalry division from the Imperial Guard under General Charles Lefebvre-Desnouettes to deal with it. On September 30, Lefebvre-Desnouettes ran into Schwarzenberg's raiders led by the Cossack Hetman Platov raiding from the south. According to Stewart, the French had the worst of this encounter, losing more of their precious light cavalry in the process.[106] Meanwhile, in the relatively quiet area of the lower Elbe near Hamburg, Clausewitz and Wallmoden destroyed a weak French division belonging to Davout on September 16 with their hodgepodge corps. Napoleon would receive no help from that quarter.[107]

By late September, Bennigsen's Army of Poland—60,000 fresh troops—had arrived in Bohemia and the time appeared ripe for an audacious coordinated move by the Sixth Coalition's armies. Schwarzenberg requested Blücher join him for a concerted drive on Leipzig. Blücher instead recommended a flank march by his army to unite with Bernadotte and Schwarzenberg endorsed Blucher's course of action.[108] In a rare spirit of cooperation, Blücher and Bernadotte both proceeded to force the line of Elbe River. Bernadotte had established a bridgehead at Rosslau in the final days of September. Blücher's appearance and crossing the Elbe took Napoleon and Ney completely by surprise. On October 3, Blücher's lead corps under General Yorck conducted a successful opposed river crossing against Bertrand's IV corps at Wartenberg further upstream (see map 33). A Prussian victory against a well-led French force did much to increase the momentum on the Allied side.[109]

Napoleon responded swiftly to this new threat. He marched on Blücher with the mass of his army, 150,000 men, leaving St. Cyr to hold Dresden with 20,000. This

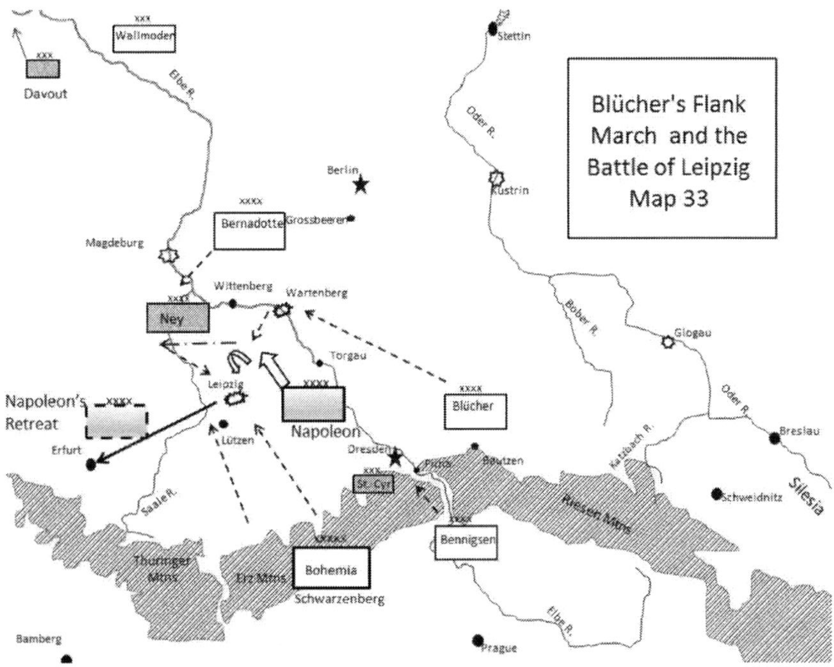

move exposed the instability of the relationship between Bernadotte and Blücher and also the general problem of synchronizing the movements of these massive armies. When Bernadotte learned of Napoleon's approach, he pulled back, out of supporting distance of Blücher. Worse, Blücher had counted on Schwarzenberg to resume his offensive to distract Napoleon. The Army of Bohemia moved slowly in the south, allowing the Emperor a chance at an operational counterstroke against the now-isolated Army of Silesia. However, timely intelligence of Napoleon's advance and excellent staff work by Gneisenau in developing a new line of operations prevented Blücher's demise. Taking a page from Napoleon's playbook, Blücher abandoned his line of communications with Berlin and dodged Napoleon's counterstroke at Duben on October 9 by retreating *to the west*. General Gneisenau wrote:

> The offensive movements of the Allied powers . . . by their execution exposed their richest provinces to . . . the enemy and by attempting to throw themselves between Napoleon and France left close in their rear an army of 200,000 men, headed by an enterprising military genius and a number of strongholds well garrisoned while they themselves had no fortified place to serve them as a rallying point or as a position to rest upon.[110]

The stage was now set for the climactic battle of Leipzig. While Napoleon attempted to trap Blücher, Schwarzenberg executed the plan that the Tsar had thwarted in August. Emboldened by the arrival of Bennigsen, the main army group advanced on Leipzig. They received the good news on October 8 that the King of Bavaria had deserted Napoleon and joined their coalition and the Austrians had pushed forward into Bavaria to join forces with them. Napoleon now had two armies across his communications: Blücher's and a combined Austro-Bavarian army under the Bavarian general Karl Philippe von Wrede. Napoleon lost the equivalent of 50,000 men with the defection of Bavaria and now realized that the 20,000 troops in Dresden were probably lost as well. Nevertheless, he concentrated his forces at Leipzig for the final battle—perhaps he could smash the Austrians before Blücher and Bernadotte arrived.

Leipzig was the largest battle in history until World War I, lasting four days and pitting Napoleon's 180,000 troops against the Allies' 200,000 (which grew by another 145,000 with the arrival of Bernadotte and Bennigsen on October 18).[111] The Allies had 1,500 cannon, over twice Napoleon's number. The battle was, in many ways, a microcosm of the entire fall campaign. All the elements were there on a tactical level—Blücher's aggressive offensive into the northern suburbs around Mockern on October 16; Bernadotte's belated advance that avoided combat the first two days of the battle; and the clash between Schwarzenberg and the Tsar over when and where to fight, finally conducting almost two distinct battles on either side of the Pleisse river, neither achieving success. The similarities were not accidental. The Allies just kept doing what they had done all along. While Schwarzenberg's army group was attacked by the bulk of Napoleon's army in the south and east, Blücher's efforts in the north, along with a direct threat to Napoleon's line of retreat by the Austrian corps of General Guylai, sealed the tactical victory by denying Napoleon the reserves he had counted on to exploit his local successes. The Army of Bohemia was pushed back, but it was not defeated. A relative lull in the battle occurred on October 17 —analogous to the period of inactivity in September after Dennewitz. The Allies were content to bring up their operational reserves—the armies of Bennigsen and Bernadotte. With the arrival of these forces Napoleon realized he was only buying time to secure his retreat.[112]

Fortune now deserted Napoleon. After a hard day's fighting on October 18 that saw the French line constrict but not break, Napoleon began a well-ordered retreat from Leipzig. At midday on October 19, the only bridge over the Elster River leading out of Leipzig to the west was prematurely blown up, cutting off three corps of the *Grande Armée*. This disaster converted the operational victory of the coalition into a strategic victory. With the Elster Bridge Napoleon lost Germany and the campaign. Napoleon's retreat would not end in Germany, but in France. He completed the campaign on a positive note when he pushed aside Wrede's Austro-Bavarian army at Hanau on October 30, gaining a path for his final withdrawal across the Rhine into

France. As at the Beresina, the victory proved expensive since many stragglers were cut off and surrendered after Napoleon's hard core of fighters, now down to around 40,000 with the colors, crossed the Rhine into France.[113] This was all that was left to Napoleon after starting the campaign two and half months earlier with over 400,000 men.

* * *

The Allies' attritional campaign had exceeded expectations, primarily due to their adherence to the spirit, if not the letter, of the Reichenbach Plan's attritional strategy and the energy and competence of the rejuvenated Prussian army as directed by Scharnhorst's "band of brothers." There was clearly evidence of operational art present during the fall operations of the armies of the Sixth Coalition in 1813. The single most compelling example was the development, promulgation, and execution of the Trachenberg Convention and subsequent Reichenbach Plan. This plan might be among the first of the modern operations orders that are so familiar to the military staffs of today. Perhaps the most arresting aspect of the plan was its brevity. Its evolution was modern in every sense of the word, combining military staff work (mostly Austrian) with a supremely political process for approval. The other astonishing aspect of the plan was its execution, which the Allies implemented implicitly throughout the campaign terminating in the operational end-state envisioned by the Trachenberg Convention—a union of all the major Allied armies in "the enemy camp" at Leipzig.

The Allied maneuvers per these plans provide evidence for a *distributed operation* where battle was subordinate to maneuver the majority of the time. The Allies executed in a distributed fashion as well per the plan (except at Dresden), in spite of command conflicts at the highest levels. The maneuvers of the armies along Napoleon's flanks and rear had the effect of engaging him throughout the depth of the battlefield. The forces of raiding Cossacks and Freikorps, too, as in the spring, had their effect throughout the depth of the battlefield that synergized with the maneuvers of the larger forces. The Allies repeatedly refused battle—in Bohemia, Silesia, and significantly at Duben south of the Elbe. Battles occurred, but their results were not expected to be decisive in the short term; Gneisenau compared it to "baiting the ravenous tiger in his den."[114]

The Allied operations also provide evidence of operational art using the criteria in chapter 1 due to their solution of the problem of command and control (C2). Instantaneous C2 was possible in those days on a broad front if intent was well understood and executed within the framework of a mission command philosophy of disciplined initiative. All of the armies, especially that of the fiery Blücher, displayed this trait throughout the campaign. It offered the ultimate in centralized planning and decentralized execution and had the effect of minimizing the negative aspects resulting from friction and fog in war, enabling success, and minimizing setbacks as long as the coalition's commanders adhered to it. On a practical level, the Prussians further enhanced C2 and unity with Gneisenau's system of chiefs of staff in the Prussian corps in the armies of Schwarzenberg and Bernadotte answering to him for the conduct of their operations.[115]

The Reichenbach Plan also provided the essential unity of effort that can be translated as evidence of *corporate operational vision*. The Allied accounts of this campaign constantly refer to actions being taken in accordance with the agreed plan of operations. Gneisenau, writing in mid-September, summarized this shared vision: "the great plan of the Allies was only ripening and the main object, of acting on the flanks of Napoleon's army, was never lost sight of."[116] Gneisenau's system within his own army empowered each corps with this vision by means of a chief of staff selected from the talented pool of officers mentored by Scharnhorst. It was no accident that Boyen was behind Bülow at Grossbeeren and Dennewitz or that Clausewitz performed the same function for Wallmoden against Davout. The wide expanse and number of these operations, along with the deployment of the major armies as specified in the plan, also provide proof of *distributed deployment*. Instead of concentrating the major forces into one huge juggernaut à la Napoleon, the Allies broke their forces up into three independent field armies. However, this deployment did cause problems, especially with the army group operating out of Bohemia. Nowhere during the fall phase of the campaign did Allied logistics play as important a role as in Schwarzenberg's area of operations. His careful planning for an advance through the Bohemian mountains toward Leipzig was frustrated by the Tsar and resulted in inadequate support during the advance to Dresden. On the northern front the supply situation was more in line with operational art because of the flow of British supplies from the Baltic ports across a less difficult terrain.

The case of the Army of Bohemia also provides more evidence in the form of *operational durability*. Despite a serious defeat at Dresden, this army group survived to turn the tables on Napoleon a few days later at Kulm. Napoleon's armies themselves also provide evidence of operational durability, suffering major defeats at the Katzbach, Grossbeeren, Dennewitz, and even Leipzig, yet fighting on—but only after the deft hand of the master was applied to help them regain their cohesion. Blücher's Army of Silesia might be the best example of all, constantly retreating and advancing, continually fighting, right up to Leipzig where Yorck's corps was fed into a meat grinder that nevertheless upset Napoleon's success against Schwarzenberg.

Finally, there was the *distributed enemy*, Napoleon and his lieutenants. Napoleon's defense was obviously meant to take advantage of interior lines using a mobile reserve and giving ground as necessary. That he was defeated doesn't mean that he was not distributed. He opposed Bernadotte with armies under Davout, Oudinot, and Ney;

Blücher with Macdonald; and left the capable St. Cyr in Dresden to watch Bohemia. He also responded to threats to his rear with significant forces while he continued to bring up soldiers that had mobilized in his rear (Augereau's corps in particular). Perhaps Napoleon's greatest operational failing was to leave so many of his veteran troops tied up holding fortresses (particularly Hamburg and Dresden) in the best 18th-century fashion as if he were fighting a limited war where cities could be traded at the peace table.

In summary, the planning and execution of the fall phase of the campaign of 1813 provides evidence that attributes of operational art were not only present, but had become more sophisticated in their manifestation. The Reichenbach Plan foreshadowed the more complex operations plans of the future. It unified and simplified command and harmony by depersonalizing the operational strategy into a document approved at a political level. Above all, it subordinated engagements to a strategy of exhaustion augmenting superior strength with maneuver. Perhaps the best evidence of the Allies' operational maturity was that their plan *did* achieve its objective, the defeat of the "enterprising genius" of Napoleon Bonaparte by attrition.

EPILOGUE

THE TWILIGHT OF NAPOLEONIC OPERATIONAL ART

FRANCE 1814 AND WATERLOO

Every campaign plan chooses one path among a thousand. . . . The mere listing of the essential facts will enable common sense—that is, a mind that is not corrupted—to identify at once what is true and right. Such a mind possesses something akin to musical sense; it easily identifies a false argument as though it were dissonant.[1]

Carl von Clausewitz, ca. 1820

Carl von Clausewitz opened his operational analysis of Napoleon's brilliant 1814 campaign with an odd allusion to a musical metaphor, but the idea of dissonance marks a convenient starting point for the end of this study. Clausewitz's translator, Peter Paret, criticizes this account's "surprising neglect of politics" and their influence on the campaign.[2] Had Napoleon been a bit better politician these campaigns, both the 1814 and Waterloo campaigns might have never occurred. It is hard to understand this failing. By the end of the 1813 campaign Napoleon was beaten and he knew it; so did every *soldat* and *l'enfant* in the *Grande Armée*—a now rather grand title for a spent force. The allies knew it, too.

However, Napoleon the statesman once again decided to see if Napoleon the general could win at the gaming table that which the statesman had squandered.[3] He rejected the Allies' initial peace terms of a return to the "natural frontiers" of France along the Rhine, Alps, and Pyrenees, and then made things much worse on himself by showing how dangerous he remained by fighting brilliantly on the defense and on interior lines as he had in Italy in 1796.[4] In doing this he hardened the Allies' collective resolve when they had begun to quibble among themselves about leaving him in power. As a result, if the quarreling Prussians, Russians, British, and Austrians could agree on anything, it was now that Napoleon should be removed for all time as the head of state of the nation of France. A similar dynamic took place when he interrupted the Congress of Vienna in March 1815 with his escape from exile in Elba by landing in France and marching on Paris. Mirroring this history, then, this book treats these final two campaigns (instead of just the Waterloo campaign) as an operational epilogue rather than to a more in-depth analysis. In any case, both of these campaigns have been analyzed at the operational level in some detail by Clausewitz, among others, and readers are encouraged to examine those analyses in all their power, keeping in mind the ideas expressed so far in this book.[5]

To a certain degree the trajectory toward the operational art went into reverse gear during the campaigns of 1814 and 1815, becoming almost 18th-century affairs when examined from the French perspective. The wars seemed to end where they started, with France surrounded on all sides and with hostile armies threatening her on all fronts, except this time there was no Valmy (see map 34). Also, the coalition he faced had a resolve hardened by the "terrible lessons" Napoleon and France had

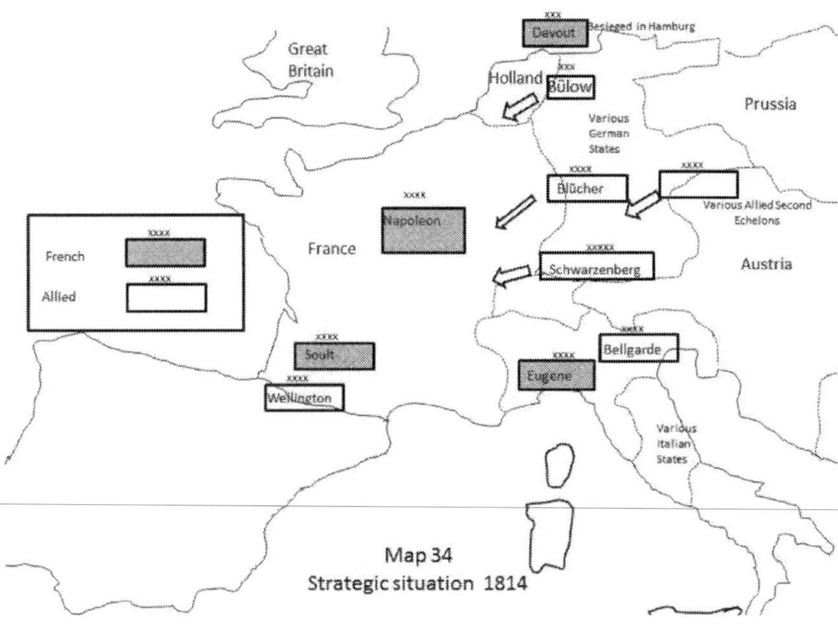

Map 34
Strategic situation 1814

administered for over 20 years.[6] In the opening days of 1814, Wellington advanced with his coalition army on a single line of operation toward Toulouse against Marshal Soult. Viceroy Eugene defended northeastern Italy in a series of rearguard actions, again along a single line of operations against the more numerous Austrians. Marshal Suchet, still undefeated, retreated reluctantly into France, leaving thousands of troops behind in garrisons that would never be relieved. As various Allied forces subdued and surrounded French fortresses, the one-time Marshal Bernadotte cashed in on his promised rewards and conquered Denmark as a means to add Norway to the Kingdom of Sweden.

Finally, Napoleon found himself penned in by several large Allied armies in the region along the Seine and the province of Champagne, campaigning in the dead of winter with an army of never more than about 80,000 troops, most of them teenage conscripts with a sprinkling of veterans and Poles who resided mostly in the Imperial Guard. On the battlefield he often only commanded half that many, and often less, as he confronted the Prussians, Russians, Austrians, and their reluctant "new" German allied contingents. Thus Napoleon was commanding essentially the same number of troops he had at the beginning of his career in 1796, in roughly the same context—conducting an operational defense of a strategic objective on interior lines. However, this time the strategic objective belonged to him and not the Austrians—Paris, not Mantua.[7]

What had changed? Ironically, the asymmetry that French arms had attained during the Revolution vis-à-vis their opponents had flip-flopped. The Allies did not fight in the 18th-century style, even though Napoleon gave them a run for their money; winning many of his battles in the winter countryside of the Seine, Marne, and Aube river valleys (see map 35). Whereas Napoleon fought with an 18th-century-style force, the Allies opposed him with army groups. The situation recalls the waning months of the American Civil War, when the tactical excellence of Lee and Johnston availed them nothing against Grant and Sherman. It was now the Allies who had the grand armies composed of hardened veterans, experienced generals, motivated by an inflamed nationalism. And they had the numbers. Against Napoleon's mid-sized army they initially advanced united with Blücher commanding nearly 60,000 troops and Schwarzenberg commanding over twice that many. Unlike Napoleon they had large operational echelons approaching the theater from the northeast numbering another 60,000 or more (principally the corps of von Bülow and Winzingerode).

The Allies made one clear operational error and Napoleon made them pay for it when they separated after defeating him early in February 1814 at the battle of La Rothiere. Napoleon, badly beaten, was inclined to accept their revised peace terms of a return to the *status quo ante bellum* of France in 1789 (peace negotiations continued throughout the period), but could not resist his good fortune when Blücher let his army get strung out on the northern line of approach to Paris. He administered

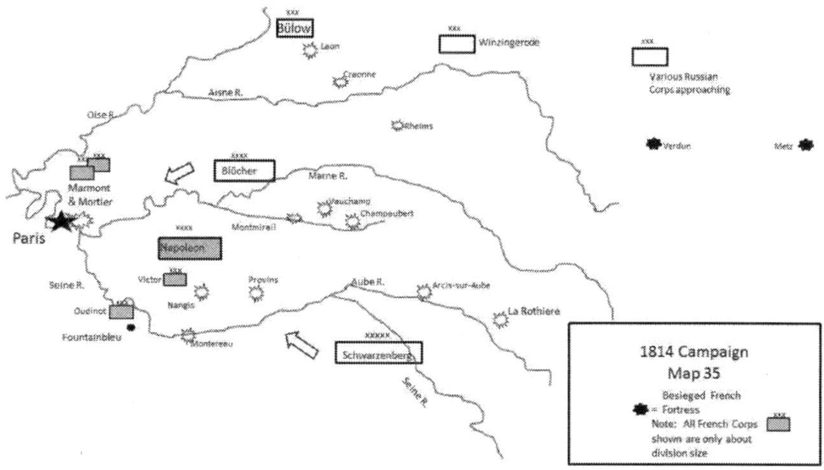

three tactical defeats against their isolated corps—at Champaubert, Montmirail, and Vauchamps (February 10–14)—which sent the Prussians reeling to the north, toward their approaching reserves. But true to form the Prussians retained their cohesion. Napoleon, as in 1813, had little time to pursue them as the large army group under Schwarzenberg (150,000) approached from the south. He managed, with a few sharp actions, to bluff it into a retreat to a safe distance south of the Seine. However, if all this reminds one of the Reichenbach-Trachenberg dynamic in 1813 with fewer Allied victories, then one is thinking along the correct lines. Napoleon's tactical victories had no meaning as long as the Allied armies remained large, operationally durable, and "in being." Worse, as mentioned, the French Emperor's tactical victories fortified their resolve to accept no peace other than one that involved his abdication.[8]

Predictably, Napoleon attempted to turn north and finish off Blücher north of the Marne River. That March, the bloody battles of Craonne (March 7) and Laon (March 9–10) showed that Napoleon could still lose as well as win, and expend his hard won advantages by further diminishing the few veteran troops available to him. Napoleon managed one more tactical victory, at Rheims, where he annihilated the corps of the émigré General Louis de St. Priest on March 13. However, his casualties had been so severe that month that he took the extraordinary expedient of proceeding east to add the garrisons of fortresses like Metz and Verdun to replace these losses. The Allied armies under Blücher and Schwarzenberg advanced in concert on Paris, ignoring Napoleon's attacks on their communications and his maneuver to the east. Napoleon had left Paris to the protection of the weak corps of Marshals Eduoard Mortier and Auguste Marmont (of Salamanca fame) and his brother Joseph. This gambit failed. Joseph, as usual, panicked; Marmont went over to the enemy with his entire corps; and Mortier rebelled along with Napoleon's other lieutenants in the so-called "revolt of the marshals." Napoleon found himself with no capital, no government, and only a small army composed mostly of his faithful guards at Fontainebleau. On April 6, he faced the inevitable and abdicated, going into exile on the small island of Elba in the Mediterranean.[9]

Napoleon's defenders often imply in their work that "if only" Joseph had not panicked, Marmont had not betrayed his old friend or the marshals had remained more resolute, that Napoleon might have prevailed in retaining his throne, or perhaps getting a regency established for his son under the Empress Marie Louise. However, these speculations ignore the facts of Russian, Prussian, and British paranoia about Austria (whose Hapsburg Princess might be regent in Paris), the claims of the Bourbons, and of numbers. Also, if the Allies could survive a Borodino or a Dresden, they could certainly survive more Vauchamps and Rheims. Napoleon, on the other hand, could not survive many more La Rothieres or Laons. His campaign to exhaust the Allies' will using the defense had failed—it failed nobly—but it had failed. There was no "miracle of the House of Bonaparte."[10] However, it impressed Clausewitz, who, it seems, decided once and for all the defense was the stronger form of war, although his inclination on this matter had been greatly supported during the 1812 campaign in Russia.[11]

SOME COMMENTS ON WATERLOO

The Waterloo Campaign in 1815 reflected the same trends noted in the previous section—the Napoleonic cycle and operational art in reverse. Although Napoleon put together something of a new *Grande Armée, L'Armée du Nord* (Army of the North), it was missing much of the original's "equipment." Marshals Lannes, Bessieres, Murat, and, perhaps especially, Berthier, were no longer there. Davout was deemed too valuable to spare and left in Paris as the minister of war. Finally, the troops themselves, many of them veterans from past wars, had little time to coalesce and train together as had the original ticket along the English Channel from 1803 until 1805 when it marched, nearly 180,000 strong, off to glory at Ulm and Austerlitz 10 years before.[12]

Another indicator of the reverse trend reflected by the 1815 campaign resides in the area of command and control. The Allied force immediately facing Napoleon had no unity of command; there was no overall army group commander. Blücher and Wellington cooperated more along the lines of Eugene of Savoy and Marlborough, operating in concert in a 19th-century version of early 18th-century military esteem. Of these two armies, Blücher's army seemed the more modern in retrospect, a veritable Prussian *Grande Armée* composed of four large combined arms corps with 30,000 men apiece. Also, he had the incomparable Gneisenau at his side as chief of staff with

Scharnhorst's acolytes (including Clausewitz) seconding all his corps commanders. Wellington's army, on the other hand, looked much like the sort of force Marlborough often took to war a hundred years earlier at slaughters like Oudenarde and Malplaquet, a coalition of disparate European contingents built around a corps of British "shock" troops, as Winston Churchill often characterized them in his famous military biography.[13] On the other hand, these two generals (three if we include Gneisenau) had proved Napoleon's most inveterate and dangerous opponents, and it did not bode well for him that the Allies had wisely placed them in close proximity to each other as their first operational echelon. However, Napoleon had little choice other than to attack them first given their proximity to Paris.

Perhaps the most positive reflection of the operational art to be found in the Waterloo campaign, at least on the French side, involved the fact that Napoleon could bring even the ghost of the *Grande Armée* with such rapidity against the Allies, themselves still mobilizing, and shatter their strategic equanimity. It is captured best in this passage from the West Point *Atlas of the Napoleonic Wars*: "This concentration remains one of the great military feats of history—all the more so in that some of these corps completed their organization as they marched."[14] However, Napoleon soon ran afoul of the frictions of war, despite the strategic surprise occasioned by his sudden arrival in mid-June 1815 at Charleroi at the hinge between the Allied armies under the Duke of Wellington and Field Marshal Blücher (see map 36). The Allies duly concentrated their individual armies prior to attempting to join forces, but on June 16 Napoleon and Marshal Ney fought Blücher and Wellington, respectively, at Ligny and Quatre Bras. As at Jena and Austerstadt, one corps wandered between both battles without participating in either, but the result was not so fortuitous. Ligny was another in a long line of almost-great Napoleonic tactical victories. However, the Prussians, though forced to retreat, did not conform to Napoleon's wish that they pull away from Wellington and maintained pace with that general and his army.[15]

Quatre Bras had not been so much a victory for the French as a stiff British rearguard action. Also, Napoleon lacked the resources of former years to pursue his enemies à outrance as he had done after Ulm, Jena, and Auerstadt. He did assign the mission of pursuing Blücher to the newest marshal Emmanuel de Grouchy, however, Grouchy's tools were limited and both he and Napoleon misjudged Blücher's line of retreat. By the time Grouchy found Blücher's rearguard under Generals von Thielmann and Clausewitz at Wavre on June 18, the other three corps of the Prussian army had already marched off toward Waterloo and Napoleon's open right flank. Meanwhile, at Waterloo (or Mont St. Jean) on June 18, Wellington fought another of his signature defensive battles—although its results looked more like Albuera than Bussaco, at least until the Prussians showed up.[16] When Blücher's first corps arrived just after 4:00 P.M., they found Wellington hard pressed but standing firm. From that

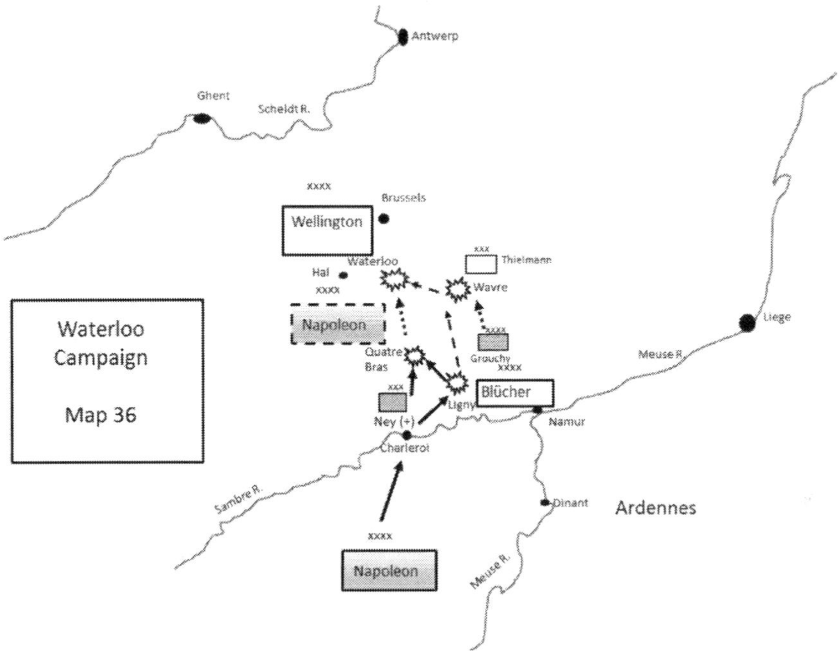

point on, Napoleon fought a losing gambit, his best chance being to shatter Wellington's army and fight Blücher's to a standstill, followed by a retreat in the manner of Leipzig with his army in relatively good order for a reunion with Grouchy. Instead, the army that was shattered was his own, thus capping his military career with his most disastrous tactical defeat.[17]

Only on the day of battle at Waterloo did the campaign shed some of its 18th-century flavor as the Allied army group, *sans* an overall commander, came together in the afternoon, increasing in size to almost 140,000 effective combatants to Napoleon's 72,000 troops (these numbers do not include deductions for casualties throughout the day). Initially Napoleon's 72,000 faced only Wellington's 68,000—although commentators have noted that Wellington might have had 85,000, which more nearly approximates the correlation of forces for the French versus the Allies at Austerlitz. Nonetheless, Blücher's message of promise to support Wellington caused that general to take the luxury of protecting his far right flank with 17,000 troops near Hal. However, with the arrival of the Prussians the campaign's reflection of the later Napoleonic operational art returns, except with the Allies better mimicking the Napoleonic method of concentration of superior force at the decisive point than the French.[18]

Comparisons with Austerlitz also fall short when these numbers and how they came to be are examined more closely. They remind one more of Napoleon at Friedland, with Napoleon in the role of Bennigsen, fighting a losing engagement as fresh echelons arrived to overwhelm him.[19] Napoleon was defeated by a mechanism that looked a lot like his own masterpieces. Blücher's army served as the *masse de manoeuvre sur la derrier* (maneuver mass on the rear) against Napoleon's flank and line of retreat. Also, it was Blücher's cavalry that executed the pursuit *à outrance* after the battle, not the French.[20]

Another comparison might be with the French maneuver at Bautzen, on a bit of a smaller scale at Waterloo, again with the roles reversed and Blücher as Ney, except that the flanking force was more effectively deployed and led. This last can be attributed to the hard work of another Prussian general staff product, Baron von Müffling. Müffling served as the often unsung liaison officer who took effective control of the deployment of Blücher's corps as they arrived on the field of battle. Jomini labels this sort of deployment "grand tactics," but in fact it is the reflection of a mature operational art on the part of the Prussians.[21] Although much ink has been spilt about "who" deserves the credit for the victory, all can surely agree that it must go to the collective, to the Allies. The point to emphasize here is that Napoleon and his army now looked like his former opponents while his opponents looked like the earlier French—the roles had reversed, both in terms of size and maneuver.

Ultimately, the campaign, as Own Connelly writes, was a "glorious irrelevance."[22] The Duke of Wellington reputedly characterized it a "social event" something "like a ball: full of interesting incidents."[23] Had Napoleon won the victory it is more than likely it would have been another of his many "pyrrhic" victories, the most recent being Ligny.[24] First, given the events of that day, in all likelihood Napoleon could not have pursued the Allies even had they been forced to retreat on divergent line from the Prussians. Wellington would have fallen back on his reserves and Blücher on his, as had happened so many times in both the 1813 and 1814 campaigns. The durable operational army created by the Napoleonic crucible now lay in the Allies' possession. Additionally, the second echelon of Allied armies was advancing on all fronts and numbered nearly a half million *more* troops, under the same resolute leaders as before: Schwarzenberg with Radetzky at his side and the ubiquitous Barclay de Tolly.[25] It was only a matter of time, not outcome, and thus Waterloo can be seen as a blessing that prevented the further effusion of needless blood. That was how it was regarded at the time, and the collapse of France so quickly afterward confirms that the spirit of 1794 was no longer there to defend *La Patrie en danger* if it meant endless war.

* * *

The twilight of Napoleonic operational art settled on Europe as military professionals tried to make sense of what happened.[26] Operational art could go only so far once the tyranny of distance and its effects on logistics and command and control limited what the new peoples' armies could do. The energies that had overcome the obstacles of 18th-century military practice were no longer inflamed by the mighty exertions that the Revolution, Napoleon, and nationalism demanded. However, as the sun set on Napoleonic glories, the tools for the next generation of operational artists were even then being forged in the diabolical workshops of men: the steam engine, the railroad, rifled artillery and muskets, and the telegraph. These workshops soon became known as factories, because of their mass production, such as Eli Whitney's famous gun factory. In them would be forged the tools for a new round of the violent development of the operational art that came to destroy the old order again—in 1914.

Appendix

THE TRACHENBERG CONVENTION

Convention signed at Trachenberg on July 12, 1813, as a basis for the operations of the campaign.

The following general principles have been decided: the Allied forces will always mass on the side of the larger enemy forces. As a consequence:

1. The corps which have to conduct operations on the enemy flanks or rear will always cut as direct as possible the enemy line of operations.
2. The larger Allied force must select a position which enables it to face the enemy wherever he advances. The salient of Bohemia seems to be proper to enable it. According to this principle, the combined armies will have to occupy the following positions before the end of the armistice.

   A part of the Allied army in Silesia (98,000 to 100,000 troops) will join as soon as possible, by the routes between Landshut and Gratz, the

Note: Translation of the Trachenberg Convention from the original French courtesy of Major Jean Parlanti, Army of the Republic of France. Charles William Vane, Marquess of Londonderry, *Narrative of the War in German and France in 1813 and 1814* (London: Henry Colburn & Richard Bentley, 1830), 373–73.

Austrian Army in order to form with it a 200,000 to 220,000 strength [army group] in Bohemia.

The army of the Crown Prince of Sweden [Bernadotte], while leaving a 15,000- to 20,000-strength [*sic*] corps screening the Danish and French from Lübeck to Hamburg, will mass approximately 70,000 troops near Trauenbrutzen. As soon as the armistice comes to an end, this army will cross the Elbe River between Torgau and Magdeburg, then move toward Leipzig.

The rest of the Allied army in Silesia, with 50,000 soldiers, will follow the enemy toward the Elbe River. This army will avoid committing itself except in the case of an extremely favorable situation. Once on the Elbe River, this force will try to cross the river between Torgau and Dresden in order to join the Crown Prince of Sweden's army. The strength will be therefore 120,000 troops. If however, there is a need, to reinforce the Allied army in Bohemia, this army instead of joining the Swedish army, will move quickly to Bohemia.

The Austrian part of the Allied force will advance either by Eger of by Hoff, or in Saxony, Silesia, or along the Danube.

If Emperor Napoleon decides to march to fight the Bohemian army, the Crown Prince of Sweden's army will try as quickly as possible to reach the enemy's rear. If, on the contrary, Napoleon moves toward the Swedish army, the Allied army will conduct a vigorous offensive operation through the enemy communications to join battle. All armies will make the enemy camp the point of rendezvous.

The Russian army (reserve) led by General Bennigsen will move from the Vistula River to the Oder by Kalisch in the direction of Glogau in order to be capable of moving toward the enemy if he stays in Silesia, or denying him the ability to invade Poland.

NOTES

PREFACE

1. For a discussion of the levels of war in U.S. military doctrine, see FM 3–0 *Operations* (February 2008) at http://downloads.army.mil/fm3–0/FM3–0.pdf (accessed December 27, 2013), 6–1, 6–2; in the recent scholarly literature a fine summation of the three levels of war can be found in Karl-Heinz Frieser, *The Blitzkrieg Legend: The 1940 Campaign in the West*, trans. and ed. John T. Greenwood (Annapolis, MD: Naval Institute Press, 2005), 6.

2. A discussion of this phenomenon by Wallace P. Franz can be found in the preface to *Two Letters on Strategy*, by Carl Von Clausewitz trans. and eds. Peter Paret and Daniel Moran (repr., Fort Leavenworth, KS: Combat Studies Institute Press, 1984), ix–xiii.

3. The author learned about the operational level of war in books by F. W. von Mellenthin and Heinz Guderian, for example, F. W. von Mellenthin, *Panzer Battles* (New York: Ballantine Books, 1971), xiii.

4. Carl von Clausewitz, *On War*, trans. and eds. Sir Michael Howard and Peter Paret (Princeton, NJ: Princeton University Press, 1976), 591–2. Unless otherwise noted, all citations refer to this translation of *On War*.

5. Macgregor Knox and Williamson Murray, eds. *The Dynamics of Military Revolution, 1300–2050* (Cambridge, UK: Cambridge University Press, 2001), 7–13. Knox and Murray define larger, "tectonic" changes in warfare as military revolutions in their introduction.

6. Thomas Paine, an Englishman, wrote *The Rights of Man* in the early 1790s for the French revolutionaries and in response to Edmund Burke's denunciation of the Revolution. Available at Project Gutenberg at http://www.ushistory.org/paine/rights/ (accessed October 24, 2013).

7. See Clausewitz, *Two Letters on Strategy*, passim, especially the introduction by Wallace Franz, ix–xii; see also Jacob W. Kipp, *Mass, Mobility, and the Red Army's Road to Operational Art, 1918–1936* (Fort Leavenworth, KS: Foreign Military Studies Office) located at: http://fmso.leavenworth.army.mil/documents/redopart.htm (accessed December 2013).

8. James J. Schneider, "Vulcan's Anvil: The American Civil War and the Emergence of the Operational Art," *Theoretical Paper Number Four* (Fort Leavenworth, KS: School of Advanced Military Studies, Command and General Staff College, 1991), 1, 38–67.

9. Jomini, Antoine Baron de. *Summary of the Art of War*, trans. Capt. G. H. Mendell and Lieut. W. P. Craighill (Westport, CT: Greenwood Press, 1992), first published in 1862 by J. B. Lippincott & Co; Clausewitz, *On War*, 128, 225.

10. David G. Chandler, *The Campaigns of Napoleon* (New York: Macmillan Publishing Co., 1966), Part 3 passim; Brent Nosworthy, *With Musket, Cannon and Sword: Battle Tactics of Napoleon and His Enemies* (New York: Di Capo Press, 1996), passim.

11. Clausewitz, *On War*, 100–12. For Clausewitz's famous characterization of Napoleon as the "God of War," see page 583.

CHAPTER 1

1. Wallace P. Franz, introduction to *Two Letters on Strategy*, by Carl Von Clausewitz, trans. and eds. Peter Paret and Daniel Moran (repr., Fort Leavenworth, KS: Combat Studies Institute Press, 1984), xiii, n2.

2. See Michael D. Krause, "Moltke and the Origins of the Operational Level of War," in *Historical Perspectives of the Operational Art*, eds. Michael D. Krause and R. Cody Philips (Washington, DC: Center of Military History, 2005), 113–14; for the linkage to Jomini, see Bruce W. Menning, "The Imperial Russian Legacy of Operational Art, 1878–1914," in *Historical Perspectives of the Operational Art*, eds. Michael Krause and R. Cody Phillips (Washington, DC: Center of Military History, U.S. Army, 2005), 189–97; for the linkage to Delbrück, see Gordon A. Craig, "Delbrück: The Military Historian," in *Makers of Modern Strategy*, ed. Peter Paret (Princeton, NJ: Princeton University Press, 1986), 326–53; and for the linkage to Clausewitz, Delbrück, and A. A. Svechin, see David Stone, "Misreading Svechin: Attrition, Annihilation, and Historicism," *The Journal of Military History* 76 (July 2012): 673–93.

3. For this characterization of the French Revolution, see Williamson Murray and MacGregor Knox, eds. *The Dynamics of Military Revolution*, introduction and chap. 4.

4. John Shy, "Jomini," in *Makers of Modern Strategy: From Machiavelli to the Nuclear Age*, ed. Peter Paret (Princeton, NJ: Princeton University Press, 1986), 146–47.

5. See Thomas M. Huber, "Jomini" in *Rise of the Western Way of War, H100 Interactive Syllabus and Book of Readings* (Fort Leavenworth, KS: Department of Military History, 2013), 222, as well as numerous conversations between the author and Dr. Huber who worked together in the Department of Military History at the Army Command and General Staff College from 2004 to 2013.

6. Anonymous, *Operations* (Washington DC: Headquarters Department of the Army, 2008), FM 3–0, Appendix A. See Clausewitz Army Doctrinal Publication (ADP) 3–0 and Army Doctrinal Reference Publication (ARDP) 3–0 that can be found here: http://armypubs.army.mil/doctrine/DR_pubs/dr_a/pdf/adrp3_0.pdf (accessed January 2015).

7. Carl von Clausewitz, *On War*, trans. and eds. Sir Michael Howard and Peter Paret (Princeton, NJ: Princeton University Press, 1976), 148.

8. Ibid., 177.

9. Ibid., FM 3–0, 6–3.

10. Ibid., 177.

11. Ibid., vi.

12. Gordon Craig, "Delbrück: The Military Historian," in *Makers of Modern Strategy: From Machiavelli to the Nuclear Age*, ed. Peter Paret (Princeton, NJ: Princeton University Press, 1984), 341–46.

13. Herbert Rosinski, "Scharnhorst to Schlieffen: The Rise and Decline of German Military Thought," *Naval War College Review* 29, no. 1 (Summer 1976): 83–103.

14. Jacob W. Kipp, "Mass, Mobility, and the Red Army's Road to Operational Art 1918–1936," accessed March 10, 2014, http://fmso.leavenworth.army.mil/documents/redopart.htm. There are no page numbers listed for this source online.

15. Bruce W. Menning, "The Imperial Russian Legacy of Operational Art," 191–92.

16. Ibid., 197–204.

17. David R. Stone, "Misreading Svechin: Attrition, Annihilation, and Historicism," *The Journal of Military History* 76 (July 2012): 673–93.

18. Cited in Kipp, "Mass and Mobility."

19. Stone, 677–78, "Misreading Svechin"; see also Kipp, "Mass and Mobility."

20. Stone, "Misreading Svechin," 690–93.

21. James J. Schneider, *The Structure of Strategic Revolution: Total War and the Roots of the Soviet Warfare State* (Novato, CA: Presidio, 1994), 3. Schneider uses the metaphor of the Soviet warfare state as a sprinter who is "always preparing to race from a running start."

22. See V. K. Triandafillov, *Nature of the Operations of Modern Armies*, trans. William A. Burhans (Moscow-Leningrad: 1929, copy located at Combined Arms Research Library, Fort Leavenworth, Kansas); and G. S. Isserson, *The Evolution of Operational Art*, trans. Bruce W. Menning (Fort Leavenworth, KS: Combat Studies Institute Press, 2013).

23. Isserson, *The Evolution of Operational Art*, 57–58.

24. James J. Schneider, "Vulcan's Anvil: The American Civil War and the Emergence of the Operational Art," *Theoretical Paper Number Four* (Fort Leavenworth, KS: School of Advanced Military Studies, Command and General Staff College, 1991), 1, 38–67.

25. Ibid., 40–41.

26. Ibid., 45–46.

27. Ibid., 54.

28. Ibid., 55–56.

29. Ibid.; for a discussion of the U.S. Army's concept of mission command, see Donald P. Wright, general ed. *16 Cases of Mission Command* (Fort Leavenworth, KS: Combat Studies Institute Press, 2013), iii–vi.

30. James J. Schneider, "Vulcan's Anvil."

31. Russell F. Weigley, *The Age of Battles* (Bloomington: Indiana University Press, 1991), xii; see also Robert M. Epstein, *Napoleon's Last Victory: 1809 and the Emergence of Modern War* (Lawrence, KS: University Press of Kansas, 1994), 173. Weigley argues that battles, in the context of winning a war in an afternoon, are indecisive. The durability of armies, particularly the nation-state army of mass conscription, is one element that contributed to this indecisiveness. Christopher Duffy's *Russia's Military Way to the West* (London: Routledge, Kegan, & Paul, 1981) provides an excellent account of the development of the Russian Army, perhaps the most durable of European Armies prior to the Napoleonic period.

32. Charles Oman, *History of the Peninsula War*, 7 vols. (Oxford: Clarendon Press, 1930), passim; for a more recent discussion, see Rory Muir, *Wellington: The Path to Victory, 1769–1814* (New Haven, CT: Yale University Press, 2013), 315.

33. Schneider, "Vulcan's Anvil," 56–59.

34. For a complete discussion of compound warfare, see Thomas M. Huber, ed., *Compound Warfare: That Fatal Knot* (Fort Leavenworth, KS: U.S. Army Command and General Staff College Press, 2002).

35. Schneider, "Vulcan's Anvil," 58–60.

36. See Kipp, "Mass, Mobility"; Stone, "Misreading Svechin"; and Menning, "Imperial Russian Legacy."

CHAPTER 2

1. Thomas M. Huber, "The Rise of Napoleon," in *H100: Rise of the Western Way of War* (Fort Leavenworth, KS: Department of Military History, 2013), 99.

2. Steven Englund, *Napoleon: A Political Life*, 178, see especially epigram mid-page.

3. Steven T. Ross, "Napoleon and Maneuver Warfare," *The Harmon Memorial Lectures in Military History, 1959–1987*, ed. Harry R. Borowski (Washington, DC: Office of Air Force History, 1988), 309–24.

4. Ross, "Napoleon and Maneuver Warfare," 309–24.

5. Steven T. Ross, "The Development of the Combat Division in Eighteenth-Century French Armies," *French Historical Studies* 4, no. 1 (Spring 1965), 84–85.

6. Ibid., 84–46; see also John Elting, *Swords around a Throne: Napoleon's Grande Armée* (New York: The Free Press, 1988), 267–80.

7. Jerome A. Greene, *The Guns of Independence: The Siege of Yorktown, 1781* (New York: Savas Beatie, 2005), 79–87.

8. Elting, *Swords around a Throne*, 124.

9. Ibid., 81–82.

10. Ibid.

11. Ross, "Development of the Combat Division," 86.

12. Ross, "Napoleon and Maneuver Warfare"; for the Gribeauval system, see Kevin F. Kiley, *Artillery of the Napoleonic Wars, 1792–1815* (Mechanicsburg, PA: Stackpole Books, 2004); Chandler, *Campaigns of Napoleon*, 5–8; 138–39.

13. Ross, "Manuever Warfare"; see also R. R. Palmer, "Frederick the Great, Guibert, Bülow: From Dynastic to National War," in *Makers of Modern Strategy*, 105–111.

14. Leo Gershoy, *The French Revolution and Napoleon* (New York: Appleton-Century-Crofts, 1933), 206–07.

15.Gershoy, *The French Revolution*, 206–207; Ross, "Development of the Combat Division," 88.

16. Elting, *Swords around a Throne*, 123–25.

17. Ross, "Development of the Combat Division," 89. YES/jtk

18. Steven T. Ross, *Quest for Victory: French Military Strategy, 1792–1799* (London, UK: A.S. Barnes and Company, 1973), 45–47.

19. Ross, "Development of the Combat Division," 87–89.

20. Ibid., 90–92.

21. Ibid., 91–93.

22. Ross, "Napoleon and Maneuver Warfare."

23. Elting, *Swords around a Throne*, 81–82.

24. Ibid., 38–39.

25. Muir, *Wellington: The Path to Victory*, 315.

26. Ross, *Quest for Victory*, 88–91.

27. John T. Kuehn, "The Reasons for the Success of the Sixth Coalition against Napoleon" (unpublished master's thesis, Fort Leavenworth, Kansas, 1997), 17.

28. Chandler, *Campaigns of Napoleon*, 56–57; 137–39.

29. Ross, *Quest for Victory*, 94–95; Chandler, *Campaigns of Napoleon*, 46–47.

30. Chandler, *Campaigns of Napoleon*, 49, quotes the operations order at length.

31. Gunther E. Rothenberg, *The Art of War in the Age of Napoleon* (Bloomington, IN: Indiana University Press, 1980), 166–67.

32. Ibid., 168.

33. John T. Kuehn, "Coalition Tactics on the Napoleonic Battlefield and Their Influence on Unity of Effort," unpublished monograph for the School of Advanced Military Studies (Fort Leavenworth, KS: 1997), 12–14; see also Rothenberg, 167–68.

34. Ross, *Quest for Victory*, 97–98. Chandler, *Campaigns of Napoleon*, 63–76. Ross lists much higher numbers for all three armies, but the numbers listed here represent actual field forces available and do not include various garrisons and remote detachments.

35. For more on John Boyd, see Grant Hammond, *John Boyd and American Security* (Washington, DC: Smithsonian Books, 2004), see also Boyd's papers, especially "Patterns of Conflict," (December 1986), 30–37, 130–35, available from Air Power Australia at http://www.ausairpower.net/APA-Boyd-Papers.html (accessed October 1, 2013).

36. Ross, *Quest for Victory*, 99; Chandler, *Campaigns of Napoleon*, 78–79.

37. Ross, *Quest for Victory*, 99.

38. Chandler, *Campaigns of Napoleon*, 84–85; Elting, *Swords around a Throne*, 163.

39. Ross, *Quest for Victory*, 98–99.

40. Chandler, *Campaigns of Napoleon*, 92–93.

41. Ross, *Quest for Victory*, 100–1.

42. Chandler, *Campaigns of Napoleon*, 199; Vincent J. Esposito and John R. Elting, *A Military History and Atlas of the Napoleonic Wars*, rev. ed. (London: Greenhill Books, 1999), narrative with map 16.

43. Esposito and Elting, *Military History and Atlas of the Napoleonic Wars*, narrative with map 17; Ross 101.

44. Esposito and Elting, *Military History and Atlas of the Napoleonic Wars*, narrative with maps 18 and 19; Ross 101–2.

45. Ross, *Quest for Victory*, 101–2; Esposito and Elting, *Military History and Atlas of the Napoleonic Wars*, narrative with map 20.

46. Ross, *Quest for Victory*, 102–3; Esposito and Elting, *Military History and Atlas of the Napoleonic Wars*, narrative with maps 23 and 24.

47. Ross, *Quest for Victory*, 103–4; Esposito and Elting, *Military History and Atlas of the Napoleonic Wars*, narrative with maps 28 and 29.

48. Ross, *Quest for Victory*, 104.

49. Esposito and Elting, *Military History and Atlas of the Napoleonic Wars*, narrative with maps 29 and 30; Chandler, *Campaigns of Napoleon*, 122; Ross, *Quest for Victory*, 104–5.

50. Chandler, *Campaigns of Napoleon*, 123–24; Esposito and Elting, *Military History and Atlas of the Napoleonic Wars*, narrative with map 30.

51. Chandler, *Campaigns of Napoleon*, 125; Esposito and Elting, *Military History and Atlas of the Napoleonic Wars*, narrative with map 31.

52. Chandler, *Campaigns of Napoleon*, 125, takes this view that war was almost guaranteed by leaving Austria a foothold in Italy with Venice.

CHAPTER 3

1. Piers Macksey, *Statesmen at War: The Strategy of Overthrow, 1798–1799* (New York: Longman Inc., 1974), 32.

2. Bruce W. Menning, "Train Hard, Fight Easy: The Legacy of A. V. Suvorov and His 'Art of Victory,'" *Air University Review* (November–December 1986), passim.

3. Christopher Duffy, *Eagles over the Alps: Suvorov in Italy and Switzerland, 1799* (Chicago, IL: The Emperor's Press, 1999), 39.

4. Menning, "Train Hard, Fight Easy"; see also Duffy, *Eagles over the Alps*, 5.

5. Gershoy, *The French Revolution and Napoleon*, 333.

6. Paul W. Schroeder, "The Collapse of the Second Coalition," *Journal of Modern History* 59 (June 1987), 250.

7. Macksey, *Statesmen at War*, 35.

8. Ibid., 10.

9. Schroeder, "Collapse of the Second Coalition," 254.

10. Benedetto Croce, *History of the Kingdom of Naples*, trans. Frances Frenaye (Chicago: University of Chicago Press, 1965), 204.

11. Marshal Macdonald, *Recollections of Marshal Macdonald*, vol. I, trans. Stephen Louis Simeon (London: Richard Bentley & Son, 1892), 216.

12. Ibid., 220.

13. Schroeder, "Collapse of the Second Coalition," 244.

14. Joseph H. Parsons, *Historical Papers upon Men and Events of Rare Interest in the Napoleonic Epoch*, vol. II (New York, 1914), 310, 313–15. In 1800, Tsar Paul I formed a league of armed neutrality against Great Britain and seized all British shipping in Russian ports.

15. Gunther E. Rothenberg, *The Art of Warfare in the Age of Napoleon* (Bloomington, IN, 1978), 196.

16. Philip Longworth, *The Art of Victory* (New York: Holt, Rinehart & Winston, 1965), 217. The specific example in Longworth is a translation of a passage in Suvorov's *Art of Victory—Nauka Pobezhda*, sometimes translated "Science of Victory" and translated by Duffy as "How to Win"; Duffy, *Eagles over the Alps*, 37n7.

17. Gunther E. Rothenberg, *Napoleon's Great Adversaries* (Bloomington, Indiana: Indiana University Press, 1982), 59; Hans Delbrück, *History of War of the Art of War*, vol. IV, trans. Walter J. Renfroe (Westport, CN, 1985), 451; Rothenberg, *Art of Warfare*, 199.

18. Rothenberg, *Art of Warfare*, 199.

19. Christopher Duffy, *Russia's Military Way to the West* (London: Routledge & Kegan Paul, 1981), 156.

20. Walter Pinter, "Russian Military Thought: The Western Model and the Shadow of Suvorov," in *Makers of Modern Strategy*, ed. Peter Paret (Princeton, NJ, 1986), 356–57.

21. Cited in Duffy, *Eagles over the Alps*, 17.

22. Menning, "Train Hard, Fight Easy," 79. See also Frederick Anthing, *Campaigns of Count Alexander Suworow Rymnikski* (London: J. Wright, 1799), xxxiii.

23. Baron Antoine Jomini, "Summary of The Art of War," in *H100: Rise of the Western Way of War, Reading H107* (Fort Leavenworth, KS: U.S. Army Command and General Staff College, June 2012), 235, 240; Menning, "Train Hard, Fight Easy," 79–80.

24. Menning, "Train Hard, Fight Easy."

25. Rothenberg, *Napoleon's Great Adversaries*, 50–53; Duffy, *Eagles over the Alps*, 23.

26. Quoted in Peter Paret, *York and the Prussian Era of Reform* (Princeton, NJ, 1966), 73–74.

27. Rothenberg, *Napoleon's Great Adversaries*, 54.

28. K. Osipov, *Suvorov*, trans. Edith Bone (London: Hutchinson & Co, 1945), 154; Schroeder, "Collapse of the Second Coalition," 250; Karl A. Roider, *Baron Thugut and Austria's Response to the French Revolution* (Princeton, NJ: Princeton University Press, 1987), 290; Rothenberg, *Art of Warfare*, 167. Suvorov's comments on the irrelevance of Vienna's orders are found in Osipov, *Suvorov*.

29. Rothenberg, *Art of Warfare*, 167.

30. Duffy, *Eagles over the Alps*, 124, 232. Parsons, *Men and Events of Rare Interest in the Napoleonic Epoch*, 289–331, passim.

31. Duffy, *Eagles over the Alps*, 89.

32. Ross, *Quest for Victory*, 214–16.

33. Ross, 214–17; Duffy, *Eagles over the Alps*, 45.

34. Ross, *Quest for Victory*, 232; Duffy, *Eagles over the Alps*, 20; Steven Ross, "The Military Strategy of the Directory: The Campaigns of 1799," *French Historical Studies* 5, no. 2 (Autumn 1967), 175.

35. Ross, "Military Strategy of the Directory," 177.

36. Ross, *Quest for Victory*, 232–34.

37. Ross, *Quest for Victory*, lists around 120,000 French in Italy; Thomas Graham, Lord Lyndenoche, *Campaigns of 1796–1799 in Germany, Italy, Switzerland, etc.* Vol. IV (London, 1812), iv–26, passim. Graham lists approximately 130,000 French in Italy, but probably includes forces in Switzerland. Figures regarding force levels and casualties are from Graham unless otherwise noted. This study is a reprint held in the rare book room at the Combined Arms Research Library, Fort Leavenworth, Kansas. The original was published circa 1800–1803. Graham's name is nowhere in the credits, but is listed in the national archives as the probable author. Graham was a liaison officer to the Austrian Army in Italy in 1796–1798, spoke French and German fluently and is known to have visited Paris during the brief peace of Amiens to examine the French records for the 1799 campaigns. These facts support Graham's authorship and therefore his work can be considered a primary source.

38. Duffy, *Eagles over the Alps*, 46.

39. Ibid., 46; Ross, *Quest for Victory*, 240.

40. Duffy, *Eagles over the Alps*, 46–47; Graham, *Campaigns of 1796–1799*, 33.

41. Ross, *Quest for Victory*, 239–240; Duffy, *Eagles over the Alps*, 47.

42. Joint Pub 3–0, VI-7.

43. Roider, *Baron Thugut*, 300–301. Duffy, *Russia's Military Way to the West*, 216.

44. Duffy, *Eagle over the Alps*, 25.

45. Longworth, *Art of Victory*, 238–40; Duffy, *Russia's Military Way to the West*, 216.

46. Cited in Osipov, *Suvorov*, 138.

47. Ibid., 139; see also Marshal Macdonald, *Recollections of Marshal Macdonald*, ed. C. Rousset, trans. S. L. Simeon (London, 1892), 252–55.

48. Duffy, *Russia's Military Way to the West*, 216; Duffy, *Eagles over the Alps*, 23; Longworth, *Art of Victory*, 240; Osipov, *Suvorov*, 139–40

49. Duffy, *Russia's Military Way to the West*, 217.

50. Duffy, *Eagles over the Alps*, 48–49.

51. Ibid., 56–59; Ross, *Quest for Victory*, 241.

52. Ross, *Quest for Victory*, 241; Duffy, *Eagles over the Alps*, 60–68.

53. Longworth, *Art of Victory*, 240; Graham, *Campaigns of 1796–1799*, 95. The account of St. Juliano is from Graham and is almost exactly the tactic discussed by Longworth.

54. Ross, *Quest for Victory*, 242.

55. Ibid., 234.

56. Ibid., 234–35.

57. Graham, *Campaigns of 1796–1799*, 129.

58. Macdonald, *Recollections of Marshal Macdonald*, 248.

59. Ibid., 252.

60. Duffy, *Russia's Military Way to the West*, 218; Macdonald, *Recollections of Marshal Macdonald*, 253; Longworth, *Art of Victory*, 251–52. Longworth also discusses Suvorov's column to double-line maneuver during this assault.

61. Macdonald, *Recollections of Marshal Macdonald*, 271; Longworth, *Art of Victory*, 252–53; Osipov, *Suvorov*, 151; Duffy, *Eagles over the Alps*, 129.

62. Duffy, *Russia's Military Way to the West*, 218; Macdonald, *Recollections of Marshal Macdonald*, 255–56; Ross, *Quest for Victory*, 246–47. The sources disagree about the losses, but it

is certain that each side lost heavily and most of the French losses seem to have come on the last two days of the battle. See Graham, *Campaigns of 1796–1799*, for details of Bellegarde's abortive attack.

63. Schroeder, "Collapse of the Second Coalition," 253 (for Austrian complaints) and 251. Schroeder discusses how Suvorov had been subject to the control of the Austrian Emperor Francis, who now delegated this responsibility to Thugut.

64. Duffy, *Eagles over the Alps*, 26.

65. Osipov, *Suvorov*, 151; Longworth, *Art of Victory*, 252–55; Duffy, *Russia's Military Way to the West*, 218; Macdonald, *Recollections of Marshal Macdonald*, 253. Osipov states: "it is a *curious* fact that Melas again completely failed to understand the situation."[emphasis original]

66. Longworth, *Art of Victory*, 255–56; Duffy, *Russia's Military Way to the West*, 216. Longworth in particular points to Bellegarde, who had been defeated by Moreau, as a principal malcontent.

67. Graham, *Campaigns of 1796–1799*, 218.

68. Osipov, *Suvorov*, 155.

69. Longworth, *Art of Victory*, 263. Duffy, *Russia's Military Way to the West*, 220. Kray's troops were in possession of a verse written by Suvorov that praised Kray.

70. Duffy, *Eagles over the Alps*, 139–48.

71. Osipov, *Suvorov*, 157.

72. Longworth, *Art of Victory*, 266. Osipov, *Suvorov*, 159.

73. Roider, *Baron Thugut*, 318–19.

74. Osipov, *Suvorov*, 159.

75. Ross, *Quest for Victory*, 235.

76. Lawrence Shadwell, translator of the Swiss narrative compiled from the writings of Archduke Charles, Baron Jomini, et al. *Mountain Warfare: Illustrated by the Campaign of 1799 in Switzerland* (London: King & Company, 1875), 144–46.

77. Ross, *Quest for Victory*, 236.

78. Ibid., 236–37; Duffy, *Eagles over the Alps*, 152–53.

79. Duffy, *Eagles over the Alps*, 160.

80. Ibid., 157, 162–63; Ross, *Quest for Victory*, 266.

81. Ross, *Quest for Victory*, 267–70.

82. Ibid., 271; Duffy, *Russia's Military Way to the West*, 219–20.

83. Ross, *Quest for Victory*, 271–72.

84. Ibid.; Duffy, *Russia's Military Way to the West*, 242–44.

85. Ross, *Quest for Victory*, 273; Duffy, *Russia's Military Way to the West*, 250, 258.

86. Macksey, *Statesmen at War*, 224.

87. Graham, *Campaigns of 1796–1799*, 264.

CHAPTER 4

1. Portions of this chapter adapted from John T Kuehn, "Warrior of the Waves: Nelson's Legacy to Naval Commandership" and from Christopher R. Gabel, and James H. Willbanks,

Great Commanders (Fort Leavenworth, Kansas: Combat Studies Institute Press, U.S. Army Combined Arms Center, 2012), 93.

2. Eric J. Grove, introduction and notes to *Some Principles of Maritime Strategy* by Julian Corbett (1911; repr., Annapolis, MD: Naval Institute Press, 1988,), 318. Sir Julian Corbett defines command of the sea in this classic work and he is, along with Captain A. T. Mahan, one of the founders of modern naval theory.

3. John Keegan, *The Price of Admiralty* (New York: Viking, 1988), 92–93.

4. Corbett, *Some Principles of Maritime Strategy*, 318.

5. See Composite Warfare Commander Concept, http://naval.dasa.ncsu.edu/sites/naval.dasa.ncsu.edu/files/Composite%20Warfare%20Concept.pdf (accessed April 27, 2014). The author operated under this doctrine for the entirety of his naval career from 1981 to 2004 and it remains the bedrock for naval operations by the U.S. Navy.

6. See Donald P. Wright, ed., *16 Cases of Mission Command* (Fort Leavenworth, KS: Combat Studies Institute Press, 2013).

7. Sun Tzu, *The Art of War*, translated with an introduction by Samuel B. Griffith (Oxford: Oxford University Press, 1963), 77.

8. The major exception to this was the performance of the French during the American Revolutionary War, especially at the Battle of the Virginia Capes; see A. T. Mahan, *The Influence of Sea Power upon History, 1660–1783* (Boston: Little Brown, 1890), 385–93, passim.

9. Admiral Richard Lord Howe was the brother of William Howe of American Revolutionary War fame.

10. In the naval terminology of Nelson's day, a "ship of the line" mounted 50 or more guns and typically 74, usually on two or more gun decks. These ships could therefore take their place in the "line" of battle and thus their name. Smaller vessels, frigates, brigs, sloops, and so on, were used for reconnaissance, screening, and patrol but often did not participate in the main battle.

11. Herbert Rosinski, *The Development of Naval Thought: Essays by Herbert Rosinski*, ed. B. Mitchell Simpson (Newport, RI: Naval War College Press, 1977), 3.

12. Rosinski, *Development of Naval Thought*, 6.

13. J. C. Wylie, *Military Strategy: A General Theory of Power Control* (New Brunswick, NJ: Rutgers University Press, 1967), 23–24; see also Lukas Milevski, "Revisiting J. C. Wylie's Dichotomy of Strategy: The Effects of Sequential and Cumulative Patters of Operations," *Journal of Strategic Studies* 35, no. 2: 239–41.

14. *Having the wind* is a naval term, sometimes called having the weather gauge. It means that the vessel A has the wind at his back (or is upwind) from vessel B (downwind). A, therefore, has the advantage and can refuse or accept battle at will and if in a faster ship can overtake the downwind vessel. Vessel A also has advantages in maneuverability if it has the wind.

15. Roger Knight, *The Pursuit of Victory: The Life and Achievement of Horatio* Nelson (New York: Basic Books, 2005), 517. This paper uses the term *battleship* to apply to ships of the line.

16. Knight, Pursuit *of Victory*, 3–7, Knight's biography portrays Nelson within the institutional context of the Royal Navy of his day.

17. Wright, *16 Cases of Mission Command*, v; Jervis's approach matches the elements in army mission command entitled: "Build cohesive teams through mutual trust" and "Create shared understanding."

18. Tom Pocock, *Horatio Nelson* (New York: Alfred A. Knopf, 1988), 121, 126–27, 345. The number in parentheses indicates the number of guns aboard the vessel. Hereafter, only the ships' names will be listed in italics without the HMS.

19. Herman, *To Rule the Waves: How the British Navy Shaped the Modern World* (New York: Harper Collins, 2004), 342–43, 345. An earlier attempt to invade Ireland had been frustrated by foul weather and a single frigate under Captain Edward Pelew of *Horatio Hornblower* fame.

20. Pocock, *Horatio Nelson,* 128–29; Herman, *To Rule the Waves,* 345–36.

21. Arthur Herman, *To Rule the Waves,* 346–49.

22. Pocock, *Horatio Nelson,* 130–33, 148; A.T. Mahan, *The Life of Nelson: The Embodiment of the Sea Power of Great Britain,* vol. 1 (Boston, MA: Little Brown and Co., 1897), 271–75.

23. Herman, *To Rule the Waves,* 151.

24. Ibid., 353; Pocock, *Horatio Nelson,* 135–45. The British lost over 150 men killed in this fiasco. The Spanish were shocked at the scale of their victory and generously returned all Nelson's wounded to him under a flag of truce.

25. Cited in Mahan, *The Life of Nelson,* 306.

26. Chandler, *The Campaigns of Napoleon,* 205–6.

27. These were Nelson's own words for his battle plan for Trafalgar, relayed to former British prime minister Henry Addington in the summer of 1805. See Colin White, *The Nelson Encyclopaedia* (London: Chatham Publishing, 2002), 236.

28. Knight, *The Pursuit of Victory,* 266–67; Herman, *To Rule the Waves,* 353.

29. Chandler, *The Campaigns of Napoleon,* 213–24.

30. Herman, *To Rule the Waves,* 353–56; Knight, *The Pursuit of Victory,* 577.

31. Herman, *To Rule the Waves,* 356–57; Captain Peter Hore, *The Habit of Victory* (London: The National Maritime Museum, 2005), 150.

32. Knight, *The Pursuit of Victory,* xxxv.

33. See Knight, *The Pursuit of Victory,* xxxi–xxxv, for a detailed description of Nelson's famous signals at the Nile. See also Chandler, Part IV for a complete discussion of Napoleon's Egyptian campaign.

34. Knight, *The Pursuit of Victory,* 288–303.

35. Ibid., 290–303; Hore, *The Habit of Victory* 150–51.

36. Knight, *The Pursuit of Victory,* xxxi–xxxv.

37. The principal cause of the war (*casus belli*) was the failure of the British to return Malta to French control as they had agreed to at Amiens. There were numerous French infractions of the treaty as well, not the least of which was Napoleon's continued territorial aggrandizements in the Netherlands, Switzerland, and Italy.

38. Herman, *To Rule the Waves,* 372–73.

39. Ibid., 376–77; the Spanish had been goaded into war when the British preemptively seized one of their treasure fleets.

40. Vincent J. Esposito and John R. Elting, *A Military History and Atlas of the Napoleonic Wars* (New York: Praeger, 1966), map 45 and facing page text. See also Knight, *The Pursuit of Victory,* 469–70.

41. Knight, *The Pursuit of Victory,* 469–70.

42. Ibid., 480.

43. Herman, *To Rule the Waves*, 376–78.

44. Knight, *The Pursuit of Victory*, 480–93; Herman, *To Rule the Waves*, 376–79.

45. Knight, *The Pursuit of Victory*, 487–94.

46. Keegan, *The Price of Admiralty*, 27–30. Hore, *The Habit of Victory*, 169. Keegan is among those who portray Calder in a poor light and argue that events surrounding Villeneuve's retreat were not strategically decisive. He makes no mention of the fact that Napoleon abandoned his invasion attempt and subsequently dispatched a replacement to relieve Villeneuve. Hore's judgment is more balanced, calling Calder's action "one of the decisive events of the whole war."

47. Knight, *The Pursuit of Victory*, 494–99.

48. See John T. Kuehn, "The Reasons for the Success of the Sixth Coalition against Napoleon in 1813," (unpublished master's thesis, Fort Leavenworth, Kansas, 1997), chapter 3, for a discussion of the formation of the Third Coalition.

49. Knight, *The Pursuit of Victory*, 501–7; Chandler, *The Campaigns of Napoleon*, 421–22.

50. Knight, *The Pursuit of Victory*, 506–7.

51. Peter Hore, *The Habit of Victory* (London: The National Maritime Museum, 2005), 178–80.

52. Knight, *The Pursuit of Victory*, 510–11.

53. Herman, *To Rule the Waves*, 386, relays that the French and Spanish certainly already felt defeated, see also Knight, *The Pursuit of Victory*, 501–7, who does not state this explicitly but implies it.

54. "Wearing" is different than "tacking" into the wind. Tacking requires better seamanship and could lead to a more confused formation if executed improperly. However, wearing was a slower, easier technique involving a gradual turn. In executing the slower maneuver Villeneuve lost valuable time in trying to flee from his pursuer. Keegan, *The Price of Admiralty*, 60.

55. Knight, *The Pursuit of Victory*, 503–17; Hore, *The Habit of Victory*, 180–82.

56. Knight, *The Pursuit of Victory*, 514.

57. Pocock, *Horatio Nelson*, 322: Knight, *The Pursuit of Victory*, 515–18.

58. Pocock, *Horatio Nelson*, 327–28.

59. Knight, *The Pursuit of Victory*, 519.

60. Ibid., 520.

61. This was the famous "Battle of the Saints." It was here that Rodney decisively defeated the French Navy under the Comte de Grasse. A.T. Mahan, *The Influence of Sea Power upon History*, chapter XIII.

62. Clark G. Reynolds, *Command of the Sea* (New York: William Morrow & Co., 1974), 307–15. See also Mark Bowden, *Napoleon's Grande Armée of 1813* (Chicago: The Emperor's Press, 1990), 26–28. For the role of the Continental System in uniting Europe against Napoleon, see John T. Kuehn, "The Reasons for the Success of the Sixth Coalition against Napoleon in 1813," master's thesis (Fort Leavenworth, KS: U.S. Army Command and General Staff College, 1997), chapter 3 passim.

CHAPTER 5

1. Cited in Steven Englund, *Napoleon: A Political Life* (Harvard University Press, 2004), 536n6.

2. Leo Tolstoy, *War and Peace*, trans. Constance Garnett (New York: The Modern Library, no date listed), 1.

3. See John T. Kuehn, "Reasons for the Success of the Sixth Coalition against Napoleon in 1813," 40–42.

4. Chandler, *Campaigns of Napoleon*, 327–28.

5. Both Jomini and Clausewitz attribute this innovation to Napoleon and emphasize in their theoretical writing. See Jomini, *Summary of the Art of War*, 244, 266; Clausewitz, *On War*, 195, 204, 208.

6. Rothenberg, *The Art of War in the Age of Napoleon*, 128; see also Chandler, *Campaigns of Napoleon*, 332–33.

7. Donald D. Horward, "Austerlitz and Masséna's Army of Italy," in *The Consortium on the revolutionary Era 1750–1850, Selected Papers 2006*, eds. F. C. Schneid and Denise Davidson (High Point University, 2007), 205–6.

8. Rothenberg, *Art of War*, 128, see also Elting, *Swords around a Throne*, 249–56.

9. Rothenberg, *Art of War*, 143–44; see Chandler, *Campaigns of Napoleon*, Appendix D, 1103 for order of battle at Austerlitz.

10. Elting, *Swords around a Throne*, 236–37; Rothenberg, *Art of War*, 141–42.

11. Chandler, *Campaigns of Napoleon*, 367–78.

12. Elting, *Swords around a Throne*, 170–73: John A. Lynn, "Toward an Army of Honor: The Moral Evolution of the French Army, 1789–1815," *French Historical Studies* 16, no. 1 (Spring 1989): 152–73.

13. Jack A. Meyer, "Napoleon's Generalship Reconsidered, or did Napoleon Really Blunder to Glory?" in *The Consortium on the Revolutionary Era 1750–1850, Selected Papers 1994*, eds. Ronald Caldwell, Donald Horward, John W. Rooney, Jr., and John Severn (Tallahassee, FL: Florida State University Press), 539–46; author conversation with Professor Connelly in 2005 revealed that Connelly preferred the title "Scrambling to Victory" but was dissuaded from using it by his publishers.

14. Jon T. Sumida, *Decoding Clausewitz: A New Approach to On War* (Lawrence, KS: University of Kansas Press, 2011), from Appendix 3 in the paperback edition, Addenda provided to John T. Kuehn via e-mail on December 29, 2010 by Jon T. Sumida.

15. Clausewitz, *On War*, 580.

16. Rothenberg, *Art of War*, 128–29; for Napoleon's semaphore system, see Chandler, *Campaigns of Napoleon*, 365–67.

17. Clausewitz, *On War*, 566. For Clausewitz the culminating point of victory was that point of diminishing returns, when the offensive had spent the power of the attacker.

18. Chandler, *Campaigns of Napoleon*, 330–32.

19. Chandler, *Campaigns of Napoleon*, 382–85, see also chapter 3.

20. Chandler, *Campaigns of Napoleon*, 390.

21. Esposito and Elting, *Military History and Atlas of the Napoleonic Wars*, map 47 and facing page text.

22. Esposito and Elting, *Military History and Atlas of the Napoleonic Wars*, map 49 and facing page text.

23. Chandler, *Campaigns of Napoleon*, 397–402.

24. Chandler, *Campaigns of Napoleon*, 402–3.

25. Chandler, *Campaigns of Napoleon*, 402–402.

26. Chandler, *Campaigns of Napoleon*, 405–8; Esposito and Elting, *Military History and Atlas of the Napoleonic Wars*, map 52 and facing page text.

27. Esposito and Elting, *Military History and Atlas of the Napoleonic Wars*, map 54 and facing page text.

28. Leslie Anders, "Austerlitz, a Clash of Command Systems," *Military Review* 38, no. 3 (June 1958): 50–57.

29. Anders, "Austerlitz" 50–57; Rothenberg, *Art of War*, 146.

30. Chandler, *Campaigns of Napoleon*, 417–33.

31. Chandler, *Campaigns of Napoleon*, 439–47; Owen Connelly, *Napoleon's Satellite Kingdoms* (New York: The Free Press, 1965), 62–64.

32. Lit. "to excess."

33. Chandler, *Campaigns of Napoleon*, 443–52, 456.

34. "Old Fritz" was the contemporary nickname of Frederick the Great.

35. Peter Paret, "The Genesis of On War," in Carl von Clausewitz, *On War*, 8; and Baron Carl von Müffling, *Memoirs of Baron Carl von Müffling*, trans. P.H. Yorke (repr., London: Greenhill Books, 1997), 5.

36. Peter Paret, ed. and trans., *Carl von Clausewitz Historical and Political Writings* (Princeton, NJ: Princeton University Press, 1992), 90–91.

37. Müffling, *Memoirs of Baron Carl von Müffling*, 5–8; Paret, *Carl von Clausewitz*, 89.

38. Rothenberg, *Art of War*, 188–89; Frederick the Great famously defeated a French army at Rossbach in 1757 as a result of an audacious cavalry attack by Friederich von Seydlitz, his cavalry commander, see Russell Weigley, *The Age of Battles* (Bloomington, IN: Indiana University Press, 1991), 184–86; Clausewitz is cited in Chandler, *Campaigns of Napoleon*, 454.

39. Rothenberg, *Art of War*, 189–90; Weigley, *Age of Battles*, 391–92; Chandler, *Campaigns of Napoleon*, 456–58. Rothenberg lists the Prussian field force at 128,000. Chandler lists the Prussian strength as 150,000, rising to 170,000 when the Saxon Army was forcibly added to its ranks in September.

40. In Paret, *Carl von Clausewitz*, 77–81.

41. David G. Chandler, "Napoleon, Operational Art, and the Jena Campaign," in *Historical Perspectives of the Operational Art*, 39–44.

42. David G. Chandler, *Jena 1806: Napoleon Destroys Prussia* (Oxford, UK: Osprey Campaign Series, 1993), 9.

43. Cited in Chandler, "Napoleon, Operational Art, and the Jena Campaign," 44.

44. Chandler, "Napoleon, Operational Art, and the Jena Campaign," 38–39.

45. Chandler, "Napoleon, Operational Art, and the Jena Campaign," 45–46.

46. Chandler, "Napoleon, Operational Art, and the Jena Campaign," 46–47; Esposito and Elting, *Military History and Atlas of the Napoleonic Wars*, map 62 and facing page text.

47. Chandler, "Napoleon, Operational Art, and the Jena Campaign," 47–48.

48. Ibid., 50–52; Esposito and Elting, *Military History and Atlas of the Napoleonic Wars*, map 64.

49. Chandler, "Napoleon, Operational Art, and the Jena Campaign," 48–58.

50. Clausewitz, *On War*, 230–35.

51. Chandler, "Napoleon, Operational Art, and the Jena Campaign," 60–61.

52. Paret, *Carl von Clausewitz*, 91–92.

53. Chandler, "Napoleon, Operational Art, and the Jena Campaign," 60–62; Elting and Esposito, map 68.

54. Chandler, "Napoleon, Operational Art, and the Jena Campaign," 62.

55. Chandler, "Napoleon, Operational Art, and the Jena Campaign," 62–63; Esposito and Elting, *Military History and Atlas of the Napoleonic Wars*, map 68 and facing page text.

56. Chandler, *Campaigns of Napoleon*, 505–09.

57. Gordon Craig, in *Makers of Modern Strategy*, 341.

58. Chandler, *Campaigns of Napoleon*, 509–12; Gregory Fremont-Barnes, ed. *The Encyclopedia of the French Revolutionary and Napoleonic Wars*, vol. II (Oxford, UK: ABC-CLIO, Inc., 2006), 413–14.

59. Esposito and Elting, *Military History and Atlas of the Napoleonic Wars*, map 68 and facing page text.

60. Ibid., map 69 and facing page text.

61. Ibid., map 71a and facing page text; see also Elting, *Swords around a Throne*, 137–38.

62. Chandler, *Campaigns of Napoleon*, 528–35.

63. Paret, *Carl von Clausewitz*, 92–93; Chandler, *Campaigns of Napoleon*, 535–55, lists only 9,000 Prussian troops actually making it to the battlefield.

64. Chandler, *Campaigns of Napoleon*, 555.

65. Esposito and Elting, *Military History and Atlas of the Napoleonic Wars*, map 75 and facing page text.

66. Chandler, *Campaigns of Napoleon*, 553–55.

67. Esposito and Elting, *Military History and Atlas of the Napoleonic Wars*, map 76 and facing page text; Chandler, *Campaigns of Napoleon*, 560–64.

68. Esposito and Elting, *Military History and Atlas of the Napoleonic Wars*, map 77 and facing page text.

69. Ibid., map 78 and facing page text.

70. Chandler, *Campaigns of Napoleon*, 586–90; Esposito and Elting, *Military History and Atlas of the Napoleonic Wars*, map 83 and facing page text.

CHAPTER 6

1. Robert M. Epstein, "Patterns of Change and Continuity in Nineteenth Century Warfare," *The Journal of Military History* 56, no. 3 (July 1992): 375–88; Thomas Huber, "Napoleon in Spain and Naples: Fortified Compound Warfare," in *Compound Warfare: That Fatal Knot* (Fort Leavenworth, KS: CGSC Press, 2002), 91–109.

2. Huber, "Compound Warfare: A Conceptual Framework," in *Compound Warfare: That Fatal Knot*, (Fort Leavenworth, KS: CGSC Press, 2002), 1–10.

3. Huber, "Napoleon in Spain and Naples: Fortified Compound Warfare," 91–109.

4. Huber, "Compound Warfare," 3.

5. Sir Charles Oman, *A History of the Peninsula War*, 7 vols. (Clarendon Press: Oxford, 1902–1930), passim. Oman makes a constant theme of Napoleon's failure to effectively exercise operational command from outside the theater.

6. Cited in Huber, "Napoleon in Spain and Naples," 91.

7. Charles Esdaille, *Napoleon's Wars: An International History* (New York: Penguin Books, 2007), 276–77.

8. Esdaille, *Napoleon's Wars*, 324; Chandler, *Campaigns of Napoleon*, 512.

9. Esdaille, *Napoleon's Wars*, 326–29.

10. Chandler, *Campaigns of Napoleon*, 596; Steven Englund, *Napoleon: A Political Life* (Cambridge, MA: Harvard University Press, 2005), 340.

11. Esdaille, *Napoleon's Wars*, 330–31.

12. Elting, *Swords around a Throne*, 36; Chandler, *Campaigns of Napoleon*, 596, 603.

13. Chandler, *Campaigns of Napoleon*, 605.

14. Esdaille, *Napoleon's Wars*, 332–45.

15. Englund, *Napoleon: A Political Life*, 340–41; Esdaille, *Napoleon's Wars*, 342–44.

16. Esposito and Elting, *Military History and Atlas of the Napoleonic Wars*, map 85 and facing page text.

17. Esposito and Elting, *Military History and Atlas of the Napoleonic Wars*, maps 84–86 and facing page text.

18. Rory Muir, *Wellington: The Path to Victory, 1796–1814* (New Haven, CT: Yale University Press, 2013), 37.

19. Muir, *Wellington: The Path to Victory*, 215–218.

20. Ibid., 229.

21. Ibid., 169–70, 235–50.

22. Ibid., 244–57; Oman, *History of the Peninsula War*, vol. 1, 260–63. Oman labels Burrard an *incubus.*

23. Oman, *History of the Peninsula War*, 279–284. Esposito and Elting, *Military History and Atlas of the Napoleonic Wars*, map 86 facing page text.

24. Oman, *History of the Peninsula War*, 291–92.

25. Esposito and Elting, *Military History and Atlas of the Napoleonic Wars*, map 87 and facing page text; Elting, *Swords around a Throne*, 361; Oman, *History of the Peninsula War*, 402–16.

26. Esposito and Elting, *Military History and Atlas of the Napoleonic Wars*, maps 90–91 and facing page text; Chandler, *Campaigns of Napoleon*, 643–657.

27. Huber, "Napoleon in Spain and Naples: Fortified Compound Warfare," 91–92.

28. Huber, "Napoleon in Spain and Naples: Fortified Compound Warfare," 92–109.

29. Cited in Muir, *Wellington: The Path to Victory*, 295.

30. Owen Connelly, *Napoleon's Satellite Kingdoms* (New York: The Free Press, 1965), 167–69. The Walcheren Expedition further reinforced the idea that Wellesley seemed to be the only British general who could win consistently against Napoleon's generals.

31. Arthur Wellesley, *The Duke of Wellington, The Dispatches of Field Marshal the Duke of Wellington during His Various Campaigns from 1799 to 1818*, vol. IV (London: John Murray, 1837), hereafter *Wellington's Dispatches*, 261.

32. *Wellington's Dispatches*, vol. 4, 261–63.

33. Muir, *Wellington: The Path to Victory*, 296–301.

34. Ibid., 301–302; Joint Publication 3–22, *Foreign Internal Defense* (Washington, DC: Department of Defense, July 2010).

35. Muir, *Wellington: The Path to Victory*, 302–9.

36. Ibid., 309–15.

37. *Wellington's Dispatches*, vol. 4, 343–45.

38. Esposito and Elting, *Military History and Atlas of the Napoleonic Wars*, map 92 and facing page text; Muir, *Wellington: The Path to Victory*, 305.

39. Muir, *Wellington: The Path to Victory*, 320–21.

40. Epstein, "Patterns of Change and Continuity," 379.

41. Muir, *Wellington: The Path to Victory*, 318–28.

42. Ibid., 329–39.

43. Ibid., 341–43.

44. Ibid., 339–44; Russell F. Weigley, *The Age of Battles*, 418–20.

45. Epstein, "Patterns of Change and Continuity," 340 and passim.

46. Rothenberg, *Art of War in the Age of Napoleon*, 179.

47. Rothenberg, *Art of War in the Age of Napoleon*, 169–70.

48. The Hungarian *Insurrectio* and Croatian *Ban* were feudal levies that were authorized in time of war and required approval by the Hungarian Diet and the Banus (governor) of Croatia. In 1809 they were theoretically capable of supplying 100,000 militia type troops for the Austrian Armies, but in fact provided only about 40,000 poorly trained and poorly equipped troops who fought for the most part in Dalmatia and Hungary. See Robert M. Epstein, *Napoleon's Last Victory* (Lawrence, KS: University of Kansas Press, 1994), chapter 3, and F. L Petre, *Napoleon and the Archduke Charles* (London: John Lane and Co., 1909), chapter 2, passim.

49. Epstein, "Patterns of Change and Continuity," 378.

50. Ibid., 377–79; Rothenberg, Art of War in the Age of Napoleon, 170–73.

51. Rothenberg, Art of War in the Age of Napoleon, especially, makes this judgment, 170; Weigley is a bit more generous, 418–21.

52. Epstein, "Patterns of Change and Continuity," 378; Weigley, *Age of Battles*, 421.

53. Epstein, "Patterns of Change and Continuity," 378; Esposito and Elting, *Military History and Atlas of the Napoleonic Wars*, maps 92–93 and facing page text.

54. Epstein, "Patterns of Change and Continuity," 378–79; Esposito and Elting, *Military History and Atlas of the Napoleonic Wars*, maps 96–100 and facing page text.

55. Esposito and Elting, *Military History and Atlas of the Napoleonic Wars*, map 100 and facing page text. When the term *French* is used henceforth it refers to Napoleon's multinational formations, too, in this case mostly Germans.

56. Weigley, *Age of Battles*, 423–24; Esposito and Elting, *Military History and Atlas of the Napoleonic Wars*, map 100 and facing page text.

57. Weigley, *Age of Battles*, 424–25.

58. Glenn J. Lamar, *Jerome Bonaparte: The War Years, 1800–1815* (Westport, CT: Greenwood Press, 2000), 61–75; Elting and Esposito, map 100 and text.

59. Weigley, *Age of Battles*, 425–29.

60. Weigley, *Age of Battles*, 428.

61. Esposito and Elting, *Military History and Atlas of the Napoleonic Wars*, map 103 and facing page text; Weigley, *Age of Battles*, 428–29.

62. Epstein, "Patterns of Change and Continuity," 379; Weigley, *Age of Battles*, 428–34.

63. Epstein, "Patterns of Change and Continuity," 379–81.

64. Felix Markham, *Napoleon* (London: The New American Library, 1963), 183.

65. Epstein, "Patterns of Change and Continuity," 378–88; Weigley, *Age of Battles*, 536–43. See chapter 1.

66. Paul Schroeder, cited in Englund, *Napoleon: A Political Life*, 324.

CHAPTER 7

1. David Bell, *The First Total War: Napoleon's Europe and the Birth of Warfare as We Know It* (New York: Houghton, Mifflin, Harcourt, 2007) advances this position; for a more modern discussion of total war as it relates to World War I, see Roger Chickering and Stig Förster, eds., *Great War, Total War* (New York: Cambridge University Press, 2000, with the German Historical Institute), 1–15.

2. Clausewitz, *On War*, Book VII: 22, 570–72.

3. Weigley, *Age of Battles*, 436.

4. Oman, *History of the Peninsula War*, vol. 3, 152; Weigley, *Age of Battles*, 436.

5. Cited in Oman, *History of the Peninsula War*, vol. 3, 210–11, 344–45; Weigley, *Age of Battles*, 435–436;

6. Oman describes this best in his introduction to volume 3 of his history, iv–v, 211.

7. Weigley, *Age of Battles*, 436.

8. Oman, *History of the Peninsula War*, vol. 3, 341–46, 353.

9. Weigley, *Age of Battles*, 436; Oman, *History of the Peninsula War*, vol. 3, 547.

10. Oman, *History of the Peninsula War*, vol. 3, 388–95, 556–58.

11. Ibid., 410–11, 437–48.

12. Weigley, *Age of Battles*, 438.

13. Ibid., 438–39.

14. Muir, *Wellington: The Path to Victory*, 418–426.

15. Weigley, *Age of Battles*, 439–40; Muir, *Wellington: The Path to Victory*, 436–45.

16. Weigley, *Age of Battles*, 486–87.

17. Ibid., 487–89.

18. Ibid., 489–90.

19. Chandler, *Campaigns of Napoleon*, 739–47; Esdaille, *Napoleon's Wars*, 449.

20. Armand de Caulaincourt, *Memoirs of General de Caulaincourt, Duke of Vicenza, 1812–1813*, ed. Jean Hanoteau, trans. Hamish Miles (London: Cassell and Company, 1935), 69.

21. Jonathan M. House, "The 1812 Campaign," in *H100 Rise of the Western Way of War, Syllabus and Book of Readings* (Fort Leavenworth, KS: Department of Military History, June 2012), 217–19.

22. Riehn, Richard K., *1812: Napoleon's Russian Campaign* (New York: McGraw-Hill, 1991), 50.

23. Chandler, *Campaigns of Napoleon*, 748–49; Riehn, *1812: Napoleon's Russian Campaign*, 29.

24. See chapter 1.

25. Esdaille, *Napoleon's Wars*, 448.

26. Riehn, *1812: Napoleon's Russian Campaign*, 50–52.

27. Chandler, *Campaigns of Napoleon*, 753–56; Riehn, *1812: Napoleon's Russian Campaign*, 425.

28. Chandler, *Campaigns of Napoleon*, 760–62; Riehn, *1812: Napoleon's Russian Campaign*, 33–36

29. See Riehn, *1812: Napoleon's Russian Campaign*, 52–53 and chapter 8 for an informative discussion of these logistical considerations.

30. Numbers come from Riehn, who lists Cossacks separately and does not include artillerists in his overall numbers in *1812: Napoleon's Russian Campaign*, 49–50.

31. Riehn, *1812: Napoleon's Russian Campaign*, 29, 88–91.

32. Fabian refers to a Roman general, Fabius Cunctator, who avoided battle with Hannibal during the Second Punic War.

33. Riehn, *1812: Napoleon's Russian Campaign*, 33–34.

34. Ibid., 33–36; Frederick the Great had created the famous entrenched camp of Bunzelwitz during the Seven Years' War, which had kept superior Austrian forces in check during a critical period.

35. Riehn, *1812: Napoleon's Russian Campaign*, 45–49; Chandler, *Campaigns of Napoleon*, 750–53, 764–65; Alexander Mikaberidze, *The Battle of Borodino: Napoleon against Kutuzov* (Barnsley, UK: Pen & Sword, 2007), 6–9, 12–13.

36. For the argument about Eugene, see Robert M. Epstein, *Prince Eugene at War: 1809* (Arlington, TX: Empire Games Press, 1984); Chandler, *Campaigns of Napoleon*, 756.

37. Esdaille, *Napoleon's Wars*, 412–14.

38. Forrest A. Miller, introduction to *The Campaign of 1812 in Russia*, by Carl von Clausewitz, trans. anonymous (reimpression of the London edition of 1843, Hattiesburg, MS: Academic International, 1970), vii.

39. Riehn, *1812: Napoleon's Russian Campaign*, 48–49; Mikaberidze, *Battle of Borodino*, 8–10.

40. Cited in Mikaberidze, *Battle of Borodino*, 9.

41. Ibid, 10; Esposito and Elting, *Military History and Atlas of the Napoleonic Wars*, map 108 and facing page text.

42. Riehn, *1812: Napoleon's Russian Campaign*, 178–83.

43. Chandler, *Campaigns of Napoleon*, 775–76.

44. Riehn, *1812: Napoleon's Russian Campaign*, 169–75; Baron de Marbot, *Memoirs of Baron de Marbot*, trans. Arthur John Butler, vol. 2 (London: Greenhill Books, 1988), 215, 290–92, 316.

45. Chandler, *Campaigns of Napoleon*, 775–77; Riehn, *1812: Napoleon's Russian Campaign*, 174–79.

46. Esposito and Elting, *Military History and Atlas of the Napoleonic Wars*, map 109 and facing page text; Riehn, *1812: Napoleon's Russian Campaign*, 181–91.

47. Riehn, *1812: Napoleon's Russian Campaign*, 184, 191.

48. Chandler, *Campaigns of Napoleon*, 777.

49. Riehn, *1812: Napoleon's Russian Campaign*, 192–98.

50. Ibid., 196–201; Esposito and Elting, *Military History and Atlas of the Napoleonic Wars*, maps 110–11 and facing page text.

51. Riehn, *1812: Napoleon's Russian Campaign*, 205.

52. Ibid., 201.

53. Mikaberidze, *Battle of Borodino*, 10–11; Riehn, *1812: Napoleon's Russian Campaign*, 268–76.

54. Clausewitz, *Campaign of 1812 in Russia*, 60; Riehn, *1812: Napoleon's Russian Campaign*, 181–90, 207–8.

55. Mikaberidze, *Battle of Borodino*, 14; Clausewitz, *Campaign of 1812 in Russia*, 113.

56. Clausewitz, *Campaign of 1812 in Russia*, 60; Mikaberidze, *Battle of Borodino*, 13–14; Riehn, *1812: Napoleon's Russian Campaign*, 204–9; Clausewitz put the number at 185,000, by this point in the campaign he was serving on the staff of Pahlen's Cavalry Corps.

57. Clausewitz, *Campaign of 1812 in Russia*, 109; Riehn, *1812: Napoleon's Russian Campaign*, 207.

58. Clausewitz, *Campaign of 1812 in Russia*, 62; Riehn, *1812: Napoleon's Russian Campaign*, 206–21.

59. Mikaberidze, *Battle of Borodino*, 14–17; Clausewitz, 62–63.

60. Clausewitz, *Campaign of 1812 in Russia*, 63–65, 142; Mikaberidze, *Battle of Borodino*, 20–22; Riehn, *1812: Napoleon's Russian Campaign*, 216.

61. Clausewitz, *Campaign of 1812 in Russia*, 64–66; Riehn, *1812: Napoleon's Russian Campaign*, 239–43; Mikaberidze, *Battle of Borodino*, 49–51, lists 31,000 militia at the battle instead of 10,000, which gave the Russians a *numerical* if not qualitative edge over Napoleon's army.

62. These numbers are contentious, but give a sense of the right proportions. See Mikaberidze, *Battle of Borodino*, 50–52, 216–18; Riehn, *1812: Napoleon's Russian Campaign*, 242–43, 255; Clausewitz, *Campaign of 1812 in Russia*, 166.

63. Riehn, *1812: Napoleon's Russian Campaign*, 254–56, 259; Mikaberidze, *Battle of Borodino*, 203–4; Clausewitz, *Campaign of 1812 in Russia*, 168–69.

64. Clausewitz, *Campaign of 1812 in Russia*, 142–43; Riehn, *1812: Napoleon's Russian Campaign*, 259–61.

65. Clausewitz, *Campaign of 1812 in Russia*, 147.

66. Chandler, *Campaigns of Napoleon*, 808; Riehn, *1812: Napoleon's Russian Campaign*, 263.

67. Clausewitz, *Campaign of 1812 in Russia*, 67–72; Adam Zamoyski, *Moscow 1812: Napoleon's Fatal March* (New York: Harper Collins, 2004), 368–75; Riehn, *1812: Napoleon's Russian Campaign*, 273–77.

68. Riehn, *1812: Napoleon's Russian Campaign*, 335–57; Clausewitz, *Campaign of 1812 in Russia*, 212.

69. Clausewitz, *Campaign of 1812 in Russia*, 211.

70. Chandler, *Campaigns of Napoleon*, 823–48; Riehn, *1812: Napoleon's Russian Campaign*, 360–90.

71. Riehn, *1812: Napoleon's Russian Campaign*, 390–97, 503; Michael Leggiere, soon to be published manuscript *Napoleon and Germany: The Franco-Prussian War of 1813: Volume II: The*

Defeat of Napoleon (Cambridge: Cambridge University Press, 2015), introduction. Leggiere lists 93,000 troops but his numbers probably included Poniatowski's V Corps of mostly new Polish recruits and Reynier's Saxons and both of these corps would be unavailable to Napoleon for the spring campaign of 1813.

72. Clausewitz, *Campaign of 1812 in Russia*, 252, 255.

CHAPTER 8

1. Clausewitz, *Campaign of 1812 in Russia*, 240–41.

2. John T. Kuehn, "The Reasons for the Success of the Sixth Coalition against Napoleon in 1813" (master's thesis, Fort Leavenworth, KS: Command and General Staff College, 1997), chapter 3 passim.

3. George F. Nafziger and Marco Gioannini, *The Defense of the Napoleonic Kingdom of Italy, 1813–1814* (Westport, CT: Praeger, 2002), 220.

4. Count von Gneisenau, *The Life and Campaigns of Field Marshal Prince Blücher*, trans. J. E. Marston (London: Sherwood, Neely & Jones, 1815), 54.

5. Weigley, *Age of Battles*, 460.

6. Rothenberg, *Art of Warfare*, 165. See also Chandler, *Campaigns of Napoleon*, 873.

7. Rothenberg, *Art of Warfare*, 188–89.

8. Carl von Clausewitz, *Historical and Political Writings*, ed. and trans. Peter Paret and Daniel Morgan (Princeton, NJ: Princeton University Press, 1992), 90; Rothenberg, *Art of Warfare*, 189. See also Walter Goerlitz, *History of the German General Staff*, trans. Brian Battershaw (New York: Praeger, 1953), 20.

9. Müffling, *Memoirs*, 4–5.

10. Rothenberg, *Art of Warfare*, 93.

11. Ibid., 191.

12. Michael Leggiere, "Bulow's 3rd Prussian Army Corps during the War of Liberation." Consortium on Revolutionary Europe 1750–1850, Selected Papers, 1996 (Tallahassee: Florida State University, 1996), 240, passim; Weigley, *Age of Battles*, 460.

13. Gneisenau, *Life and Campaigns of Field Marshal Prince Blücher*, 56.

14. Rothenberg, *Art of Warfare*, 192. Peter Hofschroer, *Prussian Reserve, Militia & Irregular Troops 1806–15* (London: Osprey Publishing, 1987), 5–7; Gneisenau, *Life and Campaigns of Field Marshal Prince Blücher*, 57–58. The *Landwehr* is not to be confused with the *Landsturm*, both of which were called up in 1813. The *Landwehr* was a "trained" militia and included many of the older *Krumper*. The *Landsturm* were an armed, untrained *levée en masse* and Gneisenau calls them by that French term.

15. Cited in Rothenberg, *Art of Warfare*, 193.

16. Gneisenau, *Life and Campaigns of Field Marshal Prince Blücher*, 59.

17. Clausewitz, *Historical and Political Writings*, 98; Chandler, *Campaigns of Napoleon*, 871–73. The addition of the *Landwehr* and other recruits would eventually result in a huge force of nearly 250,000 by August of 1813.

18. Clausewitz, *Campaign of 1812 in Russia*, 246–47; Peter Paret, *Clausewitz and the State* (Princeton, NJ: Princeton University Press, 1976), 230–31.

19. Müffling, *Memoirs*, 31.

20. U.S. Department of the Army, FM 100–5, *Operations* (Fort Monroe, VA: U.S. Government Printing Office, 1993), glossary-6.

21. Clausewitz, *Campaign of 1812 in Russia*, 213–14; Michael Leggiere, *Napoleon and Germany: The Franco-Prussian War of 1813 Volume II: The Defeat of Napoleon* (Cambridge: Cambridge University Press, 2015), introduction; Müffling, *Memoirs*, 31; Gneisenau, *Life and Campaigns of Field Marshal Prince Blücher*, 73.

22. Reihn, *1812: Napoleon's Russian Campaign*, 395–96.

23. Kuehn, "Reasons for the Success of the Sixth Coalition," 60.

24. Ibid., 63–64.

25. Hofschroer, *Prussian Reserve, Militia & Irregular Troops 1806–15*, 18–19, 58–61.

26. These figures come from both Gneisenau and Müffling.

27. Chandler, *Campaigns of Napoleon*, 874.

28. Scott Bowden, *Napoleon's Grande Armée of 1813* (Chicago: The Emperor's Press, 1990), chapters 1–4, passim.

29. Clausewitz, *Historical and Political Writing*, 98.

30. Charles William Vane Stewart, Marquess of Londonderry, *Narrative of the War in Germany and France in 1813 and 1814* (London: Henry Colburn & Richard Bentley, 1830), 54; Gneisenau, *Life and Campaigns of Field Marshal Prince Blücher*, 79; Müffling, *Memoirs*, 44. Stewart, Lord Castelreagh's half-brother had served with Wellington in Spain and was assigned as an observer with the Allied forces in the summer of 1813.

31. Stewart, *War in Germany and France in 1813 and 1814*, 20.

32. Müffling, *Memoirs*, 32.

33. Cited in Chandler, *Campaigns of Napoleon*, 878.

34. Cited in Chandler, *Campaigns of Napoleon*, 887.

35. Gneisenau, *Life and Campaigns of Field Marshal Prince Blücher*, 106.

36. Esposito and Elting, *Military History and Atlas of the Napoleonic Wars*, map 131 and facing page text; see also Bowden, *Napoleon's Grande Armée of 1813*, 100–101.

37. Bowden, *Napoleon's Grande Armée of 1813*, 106; Müffling, *Memoirs*, 35–41.

38. Chandler, *Campaigns of Napoleon*, 895–96; Christopher Bassford, "Jomini," http://www.clausewitz.com/readings/Bassford/Jomini/JOMINIX.htm#JOMINI (accessed August 22, 2014). Defenders of Jomini forget that the two witnesses—Ney and Berthier—who can best attest to his errors, both died in 1815 and could never repudiate his libels of them in the years after the wars had ended.

39. John Kuehn, *Coalition Tactics on the Napoleonic Battlefield and Their Influence on Unity of Effort*, Monograph (Fort Leavenworth, KS: Command and General Staff College, 1997), 36–38.

40. Cited in Bowden, *Napoleon's Grande Armée of 1813*, 111–12.

41. Stewart, *War in Germany and France in 1813 and 1814*, 10.

42. Ibid., 30; Peter Hofschroer, *Prussian Reserve, Militia & Irregular Troops 1806–15* (London: Osprey Publishing, 1987), 24; Paret, *Clausewitz and the State*, 226–28, 239–46. Clausewitz eventually became the legion's commander when it was absorbed into the Prussian army between April 1814 and May 1815.

43. G. F. Nafziger, "Cossack Operations in Western Germany, Spring 1813," from Consortium on Revolutionary Europe, 1750–1850, Proceedings 1992 (Tallahassee, Florida State University, 1992), 382.

44. Stewart, *War in Germany and France in 1813 and 1814*, 5.

45. Nafziger, "Cossack Operations," 381–82. Nafziger estimates that this cavalry constituted 35 percent of the *total* available French cavalry for the theater.

46. Nafziger, "Cossack Operations," 376.

47. Stewart, *War in Germany and France in 1813 and 1814*, 65–66.

48. See chapter 1.

49. Nafziger, "Cossack Operations," 381.

50. See Bowden, *Napoleon's Grande Armée of 1813*, passim.

51. Müffling, *Memoirs*, 158–59; Gneisenau, *Life and Campaigns of Field Marshal Prince Blücher*, 135–42; Clausewitz, *Historical and Political Writings*, 85–109.

52. Kuehn, *Reasons for the Success of the Sixth Coalition*, chapter 6 passim; see also Leggiere, *Napoleon and Germany*, Vol. II, assessments, passim.

53. Clausewitz, *On War*, 583.

54. Müffling, *Memoirs*, 358–59. See also Chandler, *Campaigns of Napoleon*, 918. Müffling provides an excellent précis of Napoleon's dilemma. Chandler provides the pro-Napoleon view.

55. Gneisenau, *Life and Campaigns of Field Marshal Prince Blücher*, 142–45.

56. Henry A. Kissinger, *A World Restored* (New York: Grosset & Dunlap, 1964), 72.

57. Stewart, *War in Germany and France in 1813 and 1814*, 53.

58. Kissinger, *A World Restored*, 73.

59. Leggiere, *Napoleon and Germany*, 18–21, 30.

60. Ibid., 4–6; pages hereafter cited from unpublished manuscript.

61. Müffling, *Memoirs*, 55.

62. Franklin D. Scott, *Bernadotte and the Fall of Napoleon* (Cambridge: Harvard University Press, 1935), 84.

63. Müffling, *Memoirs*, 55–56.

64. Scott, *Bernadotte and the Fall of Napoleon*, 73.

65. Müffling, *Memoirs*, 56.

66. Leggiere, *Napoleon and Germany*, assessment, passim.

67. See Appendix: The Trachenberg Convention. Bernadotte's Army of the North included various independent commands, such as Wallmoden's corps that was observing Davout and Tauenzien's Prussian landwehr corps and when these are subtracted, his strength was approximately that of Blücher. Tauenzien's corps operated as part of Bernadotte's army and is often omitted from the total strength because it was not present at Leipzig.

68. Rothenberg, *Napoleon's Great Adversaries*, 180.

69. Kuehn, "Reasons for the Success of the Sixth Coalition," 85.

70. Albert S. Britt, *The Wars of Napoleon* (West Point, NY: United States Military Academy, 1973), 198.

71. Gordon Craig, *Problems of Coalition Warfare: The Military Alliance Against Napoleon, 1813–1814*, The Harmon Memorial Lectures in Military History Number Seven (Colorado: United States Air Force Academy, 1965), 5.

72. See Appendix. Blücher's army totaled nearly 98,000 by mid-August.

73. Britt, *Wars of Napoleon*, 198.

74. General Sir Robert Wilson, *Private Diary of General Sir Robert Wilson*, vol. 2 (London: John Murray, 1861), 85.

75. Leggiere, "Bulow's 3rd Prussian Army Corps," 25–28.

76. Cited in Felix Markham, *Napoleon* (New York: Grosset & Dunlap, 1964), 206.

77. Wilson, *Private Diary of General Sir Robert Wilson*, 85.

78. Stewart, *War in Germany and France in 1813 and 1814*, 106; Wilson, *Private Diary of General Sir Robert Wilson*, 84.

79. Müffling, *Memoirs*, 56.

80. Peter Hofschroer, *The Battle of the Nations* (London: Reed International Books, 1993), 16.

81. Roger Parkinson, *Clausewitz: A Biography* (New York: Stein & Day, 1970), 231.

82. Goerlitz, *German General Staff*, 41–42.

83. Leggiere, *Napoleon and Germany*, 30; Napoleon agreed at the last moment to abandon the Grand Duchy of Warsaw, restore Prussia to its pre-1806 borders, and return Dalmatia to Austria—but the Allies wanted the status quo ante bellum prior to 1805.

84. Chandler, *Campaigns of Napoleon*, 902.

85. Britt, *Wars of Napoleon*, 206.

86. Wilson, *Private Diary of General Sir Robert Wilson*, 104.

87. Britt, *Wars of Napoleon*, 207; Wilson, 91.

88. Chandler, *Campaigns of Napoleon*, 903–12.

89. Jean V. Moreau, *Memoirs of General Moreau*, ed. and trans. John Philippart (London: A. J. Valpy, Tookes Court, 1814), 232.

90. Wilson, *Private Diary of General Sir Robert Wilson*, 100.

91. Cited in Chandler, *Campaigns of Napoleon*, 903.

92. Poultney Bigelow, *History of the German Struggle for Liberty*, vol. II (New York: Harper & Brothers, 1896), 187. Kuehn, "Reasons for the Success of the Sixth Coalition," 80.

93. Gneisenau, *Life and Campaigns of Field Marshal Prince Blücher*, 192.

94. Ibid., 192–97.

95. Müffling, *Memoirs*, 292; see also Gneisenau, *Life and Campaigns of Field Marshal Prince Blücher*, 154.

96. Müffling, *Memoirs*, 311.

97. Macdonald, *Memoirs of Marshal Macdonald*, 55–56. Macdonald claimed that Napoleon instructed him to make a "diversion," but his account of his battle plan makes clear that the attack he planned on August 26 was more than a diversion.

98. Gneisenau, *Life and Campaigns of Field Marshal Prince Blücher*, 26, 171.

99. Britt, *Wars of Napoleon*, 208.

100. Hofschroer, *Leipzig 1813* (London, UK: Osprey, 1993), 58–61.

101. Bowden, *Napoleon's Grande Armée of 1813*, 160–61.

102. Esposito and Elting, *Military History and Atlas of the Napoleonic Wars*, maps 138–139 and facing page text.

103. Wilson, 125; Bowden, 167.

104. Gneisenau, *Life and Campaigns of Field Marshal Prince Blücher*, 26, 177–78.

105. Stewart, *War in Germany and France in 1813 and 1814*, 136–37. Gneisenau, *Life and Campaigns of Field Marshal Prince Blücher*, 26, 203–4.

106. Stewart, *War in Germany and France in 1813 and 1814*, 141–42. Bowden, 166.

107. Hofschroer, *Prussian Reserve, Militia, & Irregular Troops*, 37–42.

108. Britt, *Wars of Napoleon*, 212.

109. Leggiere, *Napoleon and Germany*, chapter 10.

110. Gneisenau, *Life and Campaigns of Field Marshal Prince Blücher*, 217–18. Gneisenau's grudging admiration of Napoleon is a departure from his usual rabid anti-French rhetoric, for example, he uses "Napoleon" vice Bonaparte and calls him a "military genius."

111. Esposito and Elting, *Military History and Atlas of the Napoleonic Wars*, maps 138–39 and facing page text, numbers approximate.

112. Weigley, *Age of Battles*, 478–82; Hofschroer, *Leipzig*, 64–88.

113. Leggiere, *Napoleon and Germany*, chapter 16. Leggiere estimates that these were followed by another 20,000 "marauders," many who never returned to the colors.

114. Gneisenau, *Life and Campaigns of Field Marshal Prince Blücher*, 182.

115. Leggiere, *Napoleon, and Germany*, see for example, Gneisenau's communications throughout chapter 9.

116. Gneisenau, *Life and Campaigns of Field Marshal Prince Blücher*, 179.

EPILOGUE

1. Clausewitz, *Historical and Political Writings*, trans. and eds. Peter Paret and Daniel Moran, 207.

2. Ibid., 206–207.

3. This allusion to Napoleon's dual nature as general and statesman comes from Chandler, *The Campaigns of Napoleon*, 660.

4. Connelly, *Blundering to Glory*, 194.

5. See Clausewitz, *Historical and Political Writings*, 205–19; and Christopher Bassford, Daniel Moran, and Gregory W. Pedlow, trans. and eds., *On Waterloo: Clausewitz, Wellington and the Campaign of 1815* (Clausewitz.com, 2010), which includes significant primary source material, letters, and the correspondence of the Duke of Wellington in response to Clausewitz's work.

6. Duchesse de Reggio, *Memoirs of Marshal Oudinot Due de Reggio*, trans. Alexander Teixeira de Mattos (New York: Appleton and Company, 1897), 272. The Tsar reputedly articulated this characterization shortly after he entered Paris in 1814.

7. See chapter 2.

8. Esposito and Elting, *Military History and Atlas of the Napoleonic Wars*, map 146 and facing page text; Connelly, *Blundering to Glory*, 194–97; Gershoy, *The French Revolution and Napoleon*, 515–17.

9. Chandler, *The Campaigns of Napoleon*, 992–1004.

10. This refers to the end of the Seven Years War, when Frederick the Great was clearly losing and his inveterate opponent the Tsarina Elizabeth of Russia died. Her son Peter, slavishly

devoted to Frederick, pulled Russia immediately out of the war and saved the Hohenzollern regime, the so-called "Miracle of the House of Brandenburg." David Fraser, *Frederick the Great. King of Prussia* (London, UK: Allen Lane, 2000), 420.

11. Clausewitz, *Historical and Political Writings*, 218.

12. Connelly, *Blundering to Glory*, 201–7.

13. Weigley, *Age of Battles*, 92–98; Winston S. Churchill, *Marlborough: His Life and Times Abridged* (New York: Charles Scribner's Sons, 1966 edition), passim.

14. Esposito and Elting, *Military History and Atlas of the Napoleonic Wars*, map 157 and facing page text.

15. Connelly, *Blundering to Glory*, 207–11.

16. See chapter 7.

17. Clausewitz, *On Waterloo*, in Bassford, Moran, and Gregory, 177–78.

18. Esposito and Elting, *Military History and Atlas of the Napoleonic Wars*, maps 157–162 and facing page text; Jeremy Black, *The Battle of Waterloo* (New York: Random House, 2010), 87–88.

19. See chapter 5.

20. Bassford, Moran, and Gregory, *On Waterloo*, 187–91.

21. Jomini, "Summary of the Art of War," in *H100: Rise of the Western Way of War* (Fort Leavenworth, KS: Department of Military History, 2012), 240–42.

22. Owen Connelly, *Blundering to Glory*, 201–02.

23. Bassford, Moran, and Gregory, *On Waterloo*, 1.

24. Chandler, *The Campaigns of Napoleon*, 1042–64.

25. Esposito and Elting, *Military History and Atlas of the Napoleonic Wars*, map 156.

26. Most famously Clausewitz in *On War* and Jomini in *Art of War*.

INDEX

ABOUT THE AUTHOR

John T. Kuehn is the General William Stofft Chair for Historical Research at the U.S. Army Command and General Staff College (CGSC). He retired from the U.S. Navy in 2004 at the rank of commander after 23 years of service as a naval flight officer. He has been teaching a variety of subjects, including military history, at CGSC since 2000. He authored *Agents of Innovation* (2008) and coauthored *Eyewitness Pacific Theater* (2008) with D. M. Giangreco, as well as numerous articles and editorials and was awarded a Moncado Prize from the Society for Military History in 2011. His latest book, released by Praeger, is *A Military History of Japan: From the Age of the Samurai to the 21st Century.*

Made in the USA
Las Vegas, NV
19 September 2026

e6221070-6a66-4e8d-a509-7699b3666436R01